U.S. Army
Combat Skills Handbook

Other Lyons Press
Military Handbooks and Manuals

U.S. Air Force Search & Rescue Handbook
U.S. Army Battlefield Intelligence Handbook
U.S. Army Combat Pistol Training Manual
U.S. Army Combat Stress Control Handbook
U.S. Army Counterintelligence Handbook
U.S. Army Desert Operations Handbook
U.S. Army First Aid Manual for Soldiers
U.S. Army Fitness Training Handbook
U.S. Army Map Reading and Land Navigation Handbook
U.S. Army Reconnaissance and Surveillance Handbook
U.S. Army Survival Handbook

U.S. Army Combat Skills Handbook

DEPARTMENT OF THE ARMY

THE LYONS PRESS
GUILFORD, CONNECTICUT
AN IMPRINT OF THE GLOBE PEQUOT PRESS

The Lyons Press is an imprint of the Globe Pequot Press.

10 9 8 7 6 5 4 3 2

Printed in the United States of America

ISBN 1-59228-730-1

The Library of Congress Cataloging-in-Publication Data is available on file.

Contents

			Page
		PREFACE	vi
CHAPTER	1.	Cover, Concealment, and Camouflage	1-1
	2.	Fighting Positions	2-1
	3.	Movement	3-1
	4.	Observation	4-1
	5.	Nuclear, Biological, and Chemical Warfare	5-1
	6.	Combat Intelligence and Counterintelligence	6-1
	7.	Communications	7-1
	8.	First Aid and Personal Hygiene	8-1
APPENDIX	A.	Mines	A-1
	B.	Demolitions	B-1
	C.	Obstacles	C-1
	D.	Urban Areas	D-1
	E.	Tracking	E-1
	F.	Survival, Evasion, and Escape	F-1
	G.	Weapons and Fire Control	G-1
	H.	Field Expedient Antiarmor Devices	H-1
	I.	Range Cards	I-1
GLOSSARY			Glossary-1
REFERENCES			References-1
INDEX			Index-1

Preface

This manual is dedicated to the soldier — the key to success on the battlefield. Wars are not won by machines and weapons but by the soldiers who use them. Even the best equipped army cannot win without motivated and well-trained soldiers. If the US Army is to win the next war, its soldiers must be motivated by inspired leadership, and they must know how to do their jobs and survive on the battlefield.

This is the soldier's field manual. It tells the soldier how to perform the combat skills needed to survive on the battlefield. These are basic skills that must be learned by soldiers in all military occupational specialties (MOS).

The words "he," "him," "his," "man," and "men," when used in this publication, represent both the masculine and feminine genders, unless otherwise specifically stated.

CHAPTER 1

Cover, Concealment, and Camouflage

GENERAL

If the enemy can see you, he can hit you with his fire. So you must be concealed from enemy observation and have cover from enemy fire.

When the terrain does not provide natural cover and concealment, you must prepare your cover and use natural and man-made materials to camouflage yourself, your equipment, and your position. This chapter provides guidance on the preparation and use of cover, concealment, and camouflage.

CONTENTS

GENERAL 1-1
COVER 1-2
CONCEALMENT 1-3
CAMOUFLAGE 1-3

COVER

Cover gives protection from bullets, fragments of exploding rounds, flame, nuclear effects, and biological and chemical agents. Cover can also conceal you from enemy observation. Cover can be natural or man-made.

TYPES OF COVER

Natural cover includes such things as logs, trees, stumps, ravines, and hollows. Man-made cover includes such things as fighting positions, trenches, walls, rubble, and craters. Even the smallest depression or fold in the ground can give some cover. Look for and use every bit of cover the terrain offers.

In combat, you need protection from enemy direct and indirect fire.

To get this protection in the defense, build a fighting position (man-made cover) to add to the natural cover afforded by the terrain.

FIGHTING POSITION WITH COVER

To get protection from enemy fire in the offense or when moving, use routes that put cover between you and the places where the enemy is known or thought to be. Use ravines, gullies, hills, wooded areas, walls, and other cover to keep the enemy from seeing and firing at you. Avoid open areas, and do not skyline yourself on hilltops and ridges.

TROOPS MOVING ALONG A RAVINE

CONCEALMENT

Concealment is anything that hides you from enemy observation. Concealment does not protect you from enemy fire. Do not think that you are protected from the enemy's fire just because you are concealed. Concealment, like cover, can also be natural or man-made.

Natural concealment includes such things as bushes, grass, trees, and shadows. If possible, natural concealment should not be disturbed. Man-made concealment includes such things as battle-dress uniforms, camouflage nets, face paint, and natural material that has been moved from its original location. Man-made concealment must blend into the natural concealment provided by the terrain.

Light discipline, noise discipline, movement discipline, and the use of camouflage contribute to concealment. Light discipline is controlling the use of lights at night by such things as not smoking in the open, not walking around with a flashlight on, and not using vehicle headlights. Noise discipline is taking action to deflect sounds generated by your unit (such as operating equipment) away from the enemy and, when possible, using methods to communicate that do not generate sounds (arm-and-hand signals). Movement discipline is such things as not moving about fighting positions unless necessary, and not moving on routes that lack cover and concealment. In the defense, build a well-camouflaged fighting position and avoid moving about. In the offense, conceal yourself and your equipment with camouflage and move in woods or on terrain that gives concealment. Darkness cannot hide you from enemy observation in either offense or defense. The enemy's night vision devices and other detection means let them find you in both daylight and darkness.

CAMOUFLAGE

Camouflage is anything you use to keep yourself, your equipment, and your position from looking like what they are. Both natural and man-made material can be used for camouflage.

Change and improve your camouflage often. The time between changes and improvements depends on the weather and on the material used. Natural camouflage will often die, fade, or otherwise lose its effectiveness. Likewise, man-made camouflage may wear off or fade. When those things happen, you and your equipment or position may not blend with the surroundings. That may make it easy for the enemy to spot you.

CAMOUFLAGE CONSIDERATIONS

Movement draws attention. When you give arm-and-hand signals or walk about your position, your movement can be seen by the naked eye at long ranges. In the defense, stay low and move only when necessary. In the offense, move only on covered and concealed routes.

Positions must not be where the enemy expects to find them. Build positions on the side of a hill, away from road junctions or lone buildings, and in covered and concealed places. Avoid open areas.

POSITION IN COVER AND CONCEALMENT ON A HILLSIDE

Outlines and shadows may reveal your position or equipment to air or ground observers. Outlines and shadows can be broken up with camouflage. When moving, stay in the shadows when possible.

SOLDIERS WITH LEADER IN A SHADOW OF A TREE

Shine may also attract the enemy's attention. In the dark, it may be a light such as a burning cigarette or flashlight. In daylight, it can be reflected light from polished surfaces such as shiny mess gear, a worn helmet, a windshield, a watch crystal and band, or exposed skin. A light, or its reflection, from a position may help the enemy detect the position. To reduce shine, cover your skin with clothing and face paint. However, in a nuclear attack, darkly painted skin can absorb more thermal energy and may burn more readily than bare skin. Also, dull the surfaces of equipment and vehicles with paint, mud, or some type of camouflage material.

TWO SOLDIERS CAMOUFLAGE EACH OTHER

Shape is outline or form. The shape of a helmet is easily recognized. A human body is also easily recognized. Use camouflage and concealment to breakup shapes and blend them with their surroundings. Be careful not to overdo it.

HELMET CAMOUFLAGED

The **colors** of your skin, uniform, and equipment may help the enemy detect you if the colors contrast with the background. For example, a green uniform will contrast with snow-covered terrain. Camouflage yourself and your equipment to blend with the surroundings.

SOLDIER IN ARCTIC CAMOUFLAGE

Dispersion is the spreading of men, vehicles, and equipment over a wide area. It is usually easier for the enemy to detect soldiers when they are bunched. So, spread out. The distance between you and your fellow soldier will vary with the terrain, degree of visibility, and enemy situation. Distances will normally be set by unit leaders or by a unit's standing operating procedure (SOP).

FIRE TEAM DISPERSED

HOW TO CAMOUFLAGE

Before camouflaging, study the terrain and vegetation of the area in which you are operating. Then pick and use the camouflage material that best blends with that area. When moving from one area to another, change camouflage as needed to blend with the surroundings. Take grass, leaves, brush, and other material from your location and apply it to your uniform and equipment and put face paint on your skin.

CAMOUFLAGED SOLDIERS

Fighting Positions. When building a fighting position, camouflage it and the dirt taken from it. Camouflage the dirt used as frontal, flank, rear, and overhead cover. Also camouflage the bottom of the hole to prevent detection from the air. If necessary, take excess dirt away from the position (to the rear).

Do not overcamouflage. Too much camouflage material may actually disclose a position. Get your camouflage material from a wide area. An area stripped of all or most of its vegetation may draw attention. Do not wait until the position is complete to camouflage it. Camouflage the position as you build.

CAMOUFLAGED FIGHTING POSITION BEING IMPROVED

Do not leave shiny or light-colored objects lying about. Hide mess kits, mirrors, food containers, and white underwear and towels. Do not remove your shirt in the open. Your skin may shine and be seen. Never use fires where there is a chance that the flame will be seen or the smoke will be smelled by the enemy. Also, cover up tracks and other signs of movement.

USING A TREE LIMB TO COVER UP A TRAIL

CAMOUFLAGED HELMETS

When camouflage is complete, inspect the position from the enemy's side. This should be done from about 35 meters forward of the position. Then check the camouflage periodically to see that it stays natural-looking and conceals the position. When the camouflage becomes ineffective, change and improve it.

Helmets. Camouflage your helmet with the issue helmet cover or make a cover of cloth or burlap that is colored to blend with the terrain. The cover should fit loosely with the flaps folded under the helmet or left hanging. The hanging flaps may break up the helmet outline. Leaves, grass, or sticks can also be attached to the cover. Use camouflage bands, strings, burlap strips, or rubber bands to hold those in place. If there is no material for a helmet cover, disguise and dull helmet surface with irregular patterns of paint or mud.

Uniforms. Most uniforms come already camouflaged. However, it may be necessary to add more camouflage to make the uniform blend better with the surroundings. To do this, put mud on the uniform or attach leaves, grass, or small branches to it. Too much camouflage, however, may draw attention.

When operating on snow-covered ground, wear overwhites (if issued) to help blend with the snow. If overwhites are not issued, use white cloth, such as white bedsheets, to get the same effect.

Skin. Exposed skin reflects light and may draw the enemy's attention. Even very dark skin, because of its natural oil, will reflect light. Use the following methods when applying camouflage face paint to camouflage the skin.

COLORS USED IN CAMOUFLAGE

	SKIN COLOR	SHINE AREAS	SHADOW AREAS
CAMOUFLAGE MATERIAL	LIGHT OR DARK	FOREHEAD, CHEEKBONES, EARS, NOSE AND CHIN	AROUND EYES, UNDER NOSE, AND UNDER CHIN
LOAM AND LIGHT GREEN STICK	ALL TROOPS USE IN AREAS WITH GREEN VEGETATION	USE LOAM	USE LIGHT GREEN
SAND AND LIGHT GREEN STICK	ALL TROOPS USE IN AREAS LACKING GREEN VEGETATION	USE LIGHT GREEN	USE SAND
LOAM AND WHITE	ALL TROOPS USE ONLY IN SNOW-COVERED TERRAIN	USE LOAM	USE WHITE
BURNT CORK, BARK CHARCOAL, OR LAMP BLACK	ALL TROOPS, IF CAMOUFLAGE STICKS NOT AVAILABLE	USE	DO NOT USE
LIGHT-COLOR MUD	ALL TROOPS, IF CAMOUFLAGE STICKS NOT AVAILABLE	DO NOT USE	USE

When applying camouflage stick to your skin, work with a buddy (in pairs) and help each other. Apply a two-color combination of camouflage stick in an irregular pattern. Paint shiny areas (forehead, cheekbones, nose, ears, and chin) with a dark color. Paint shadow areas (around the eyes, under the nose, and under the chin) with a light color. In addition to the face, paint the exposed skin on the back of the neck, arms, and hands. Palms of hands are not normally camouflaged if arm-and-hand signals are to be used. Remove all jewelry to further reduce shine or reflection.

When camouflage sticks are not issued, use burnt cork, bark, charcoal, lamp black, or light-colored mud.

CHAPTER 2

Fighting Positions

GENERAL

When defending, or when temporarily halted while making an attack, you must seek cover from fire and concealment from observation. Cover and concealment are best provided by some type of fighting position. This may be an existing hole, a hastily dug prone shelter, or a well-prepared position with overhead cover. The time available for preparation determines how well you build your position.

Your fighting position **must:**

- **Allow you to fire.**
- **Protect you from observation and direct and indirect fire.**

CONTENTS

GENERAL 2-1
COVER 2-2
CONCEALMENT 2-5
SECTORS AND FIELDS OF FIRE 2-5
HOW TO BUILD FIGHTING POSITIONS 2-8

COVER

The cover of your fighting position must be strong enough to protect you from small arms fire, indirect fire fragments, and the blast wave of nuclear explosions. The position should have frontal cover to give protection from small arms fire from the front. Natural frontal cover (trees, rocks, logs, and rubble) is best, because it is hard for the enemy to detect a position that is concealed by natural cover. If natural cover is not available, use the dirt taken from the hole you dig to build additional cover. The cover can be improved by putting the dirt in sandbags and then wetting them.

COVER

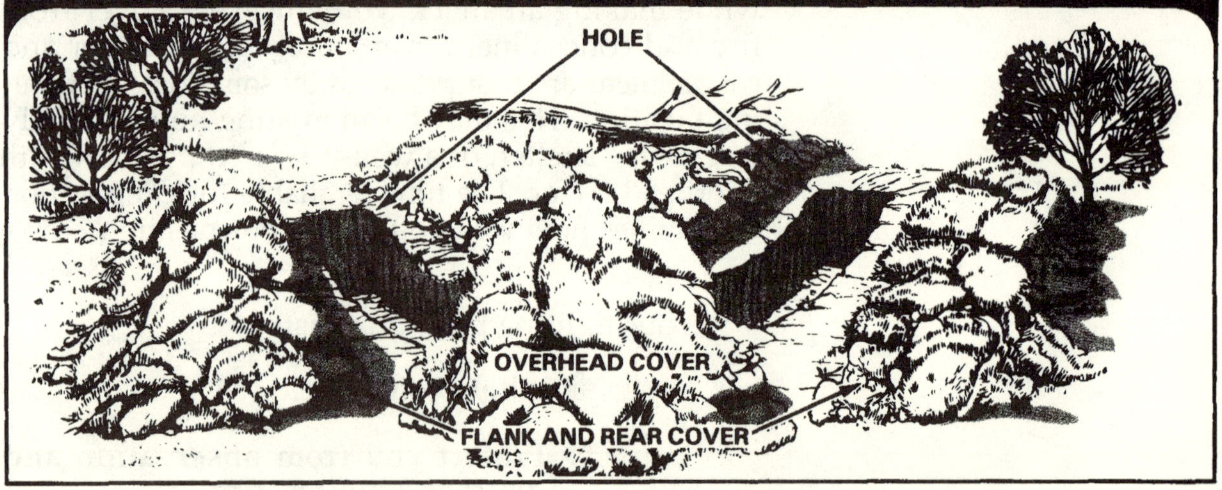

Frontal cover **must be:**

- **Thick enough (at least 46 cm [18 in] of dirt) to stop small arms fire.**
- **High enough to protect your head when you fire from behind the cover.**
- **Far enough in front of the hole to allow room for elbow holes and sector stakes so that you can fire to the oblique.**
- **Long enough to give you cover and hide the muzzle blast of your rifle when you fire to the oblique.**

Your fighting position should be built so that, when you come under direct fire from your front, you can move behind the frontal cover for protection and yet fire to the oblique.

FIGHTING POSITION

For all-round protection, to include protection from a nuclear attack, your position should also have overhead, flank, and rear cover. The dirt from the hole can also be used to build that cover, which protects against indirect fire that bursts overhead or to the flanks and rear of the position. Cover also guards against the effects of friendly weapons supporting from the rear, such as small arms fire or discarding sabot rounds fired from tanks. You should leave crawl spaces in the rear cover. This lets you enter and leave the position without exposing yourself to the enemy.

To increase your chances of survival in a nuclear attack, you should insure that your fighting position incorporates the **following considerations:**

- **Rounded walls hold up better against a blast wave than square or rectangular walls, and rounded walls are easier to dig.**
- **Small openings help keep out radiation. Most nuclear radiation in the bottom of the position is scattered into the position through the opening.**
- **Deeper fighting positions place a greater thickness of shielding material or earth between you and the nuclear detonation therefore, deep positions provide greater reduction of initial radiation entering the hole. Radiation is reduced by a factor of two for each 16 inches of hole depth.**
- **Low body positions put more dirt between you and the source of radiation. Curling upon your side or, better yet, lying on your back with knees drawn up to the chest is best. Tucked-up legs and arms tend to shield the body from radiation.**
- **Thermal radiation enters your fighting position by line of sight or by reflection off the sides. Dark and rough materials (such as wool blankets and shelter halves) can be used to cover potential reflecting surfaces.**

COMPLETED POSITION WITH ALL-ROUND COVER AND CRAWL SPACES

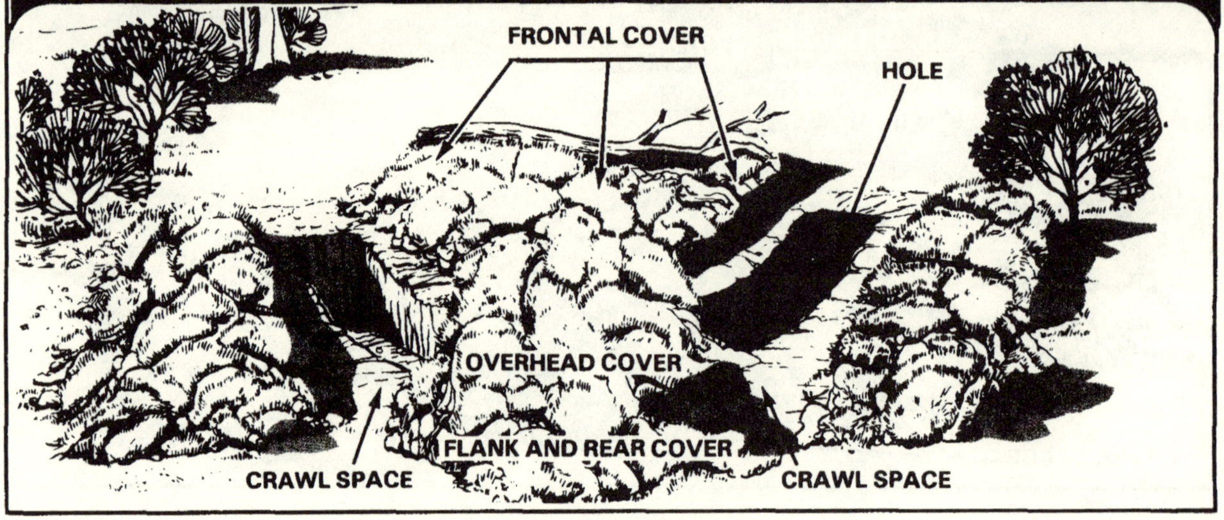

Your fighting position should be built so that, when you come under direct fire from your front, you can move behind the frontal cover for protection and yet fire to the oblique.

FIGHTING POSITION WITH SECTOR OF FIRE SHADED

To be able to see and fire into your sectors of fire, you may have to clear some vegetation and other obstructions from them. That is called clearing a field of fire.

When clearing a **field of fire:**

- **Do not disclose your position by careless or too much clearing.**
- **Leave a thin, natural screen of vegetation to hide your position.**
- **Cut off lower branches of large, scattered trees in sparsely wooded areas.**
- **Clear underbrush only where it blocks your view.**
- **Remove cut brush, limbs, and weeds so the enemy will not spot them.**
- **Cover cuts on trees and bushes forward of your position with mud, dirt, or snow.**
- **Leave no trails as clues for the enemy.**

SOLDIERS CLEARING FIELDS OF FIRE

A field of fire **to the front** is needed out to the range of your weapon.

SOLDIERS FIRING TO THE FRONT

A field of fire **to the oblique** lets you hit the attackers from an unexpected angle. It also lets you support the positions next to you. When fired to the oblique, your fire interlocks with the fire of other positions. That helps create a wall of fire that the enemy must pass through.

SOLDIERS FIRING TO THE OBLIQUE WHILE UNDER FIRE

HOW TO BUILD FIGHTING POSITIONS

HASTY FIGHTING POSITION

When there is little time for preparation, build a hasty fighting position. It should be behind whatever cover is available. It should give frontal cover from enemy direct fire but allow firing to the front and the oblique. The term **hasty** does not mean that there is no digging.

If there is a natural hole or ditch available, use it. If not, dig a prone shelter that will give some protection. The hole should be about one-half meter (18 to 20 in) deep. Use the dirt from the hole to build cover around the edge of the position.

HASTY POSITION

TWO-MAN FIGHTING POSITION

In the defense, you and another soldier will build a two-man fighting position. Improve your position as time permits.

Keep the hole small. The smaller the hole, the less likely it is that rounds, grenades, or airburst fragments will get into it. It should be large enough for you and your buddy in full combat gear. It should extend beyond the edges of the frontal cover enough to let you and your buddy observe and fire to the front. The hole is usually dug straight, but it may be curved around the frontal cover.

Curving the hole around the frontal cover may be necessary in close terrain to allow better observation and fire to the front and to the next flank position. To curve the hole, simply extend one or both ends of it around the frontal cover.

STRAIGHT AND CURVED HOLES

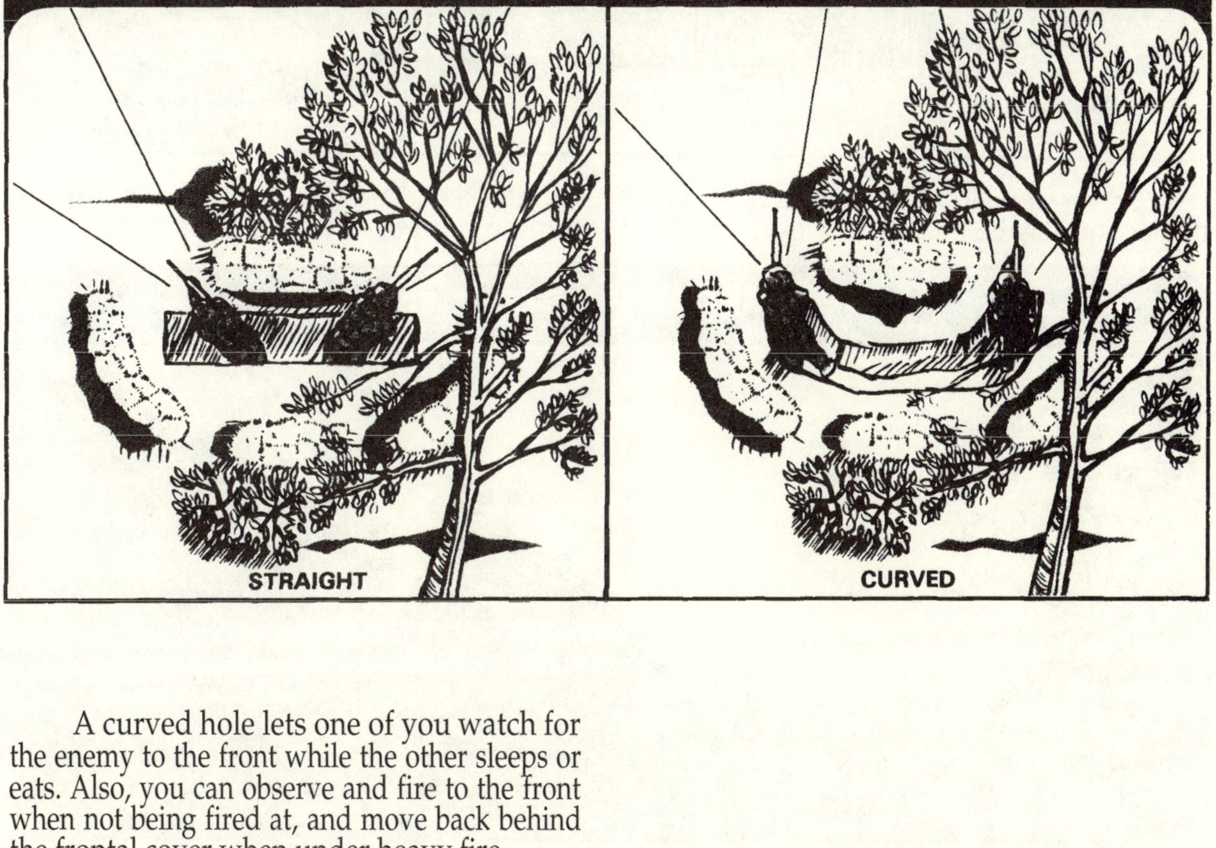

A curved hole lets one of you watch for the enemy to the front while the other sleeps or eats. Also, you can observe and fire to the front when not being fired at, and move back behind the frontal cover when under heavy fire.

MOVING BACK TO FRONT

On a steep slope, a straight hole may not let you stay behind frontal cover and fire at attackers. You may have to stand up and expose yourself to the attackers' fire.

THE EFFECT OF A STEEP SLOPE

To avoid such exposure, dig firing ports in each end of the hole. The ground between the firing ports will then be additional frontal cover.

Dig the hole **armpit deep.** This lowers your profile and still lets you fire. Other dimensions should be the length of two M16s and the width of two bayonets.

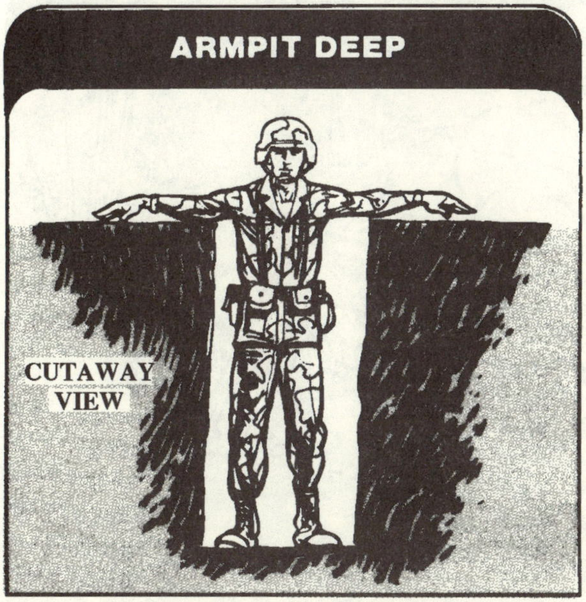

Leave enough distance between the hole and the frontal cover to make a shelf where you can put your elbows when firing.

ELBOW HOLES

Dig **elbow holes** to keep your elbows from moving around when you fire. Your fire will then be more accurate.

If you or your buddy has an automatic rifle, dig a small **trench** to stabilize its bipod legs.

TRENCH FOR BIPOD LEGS

Hammer in **sector stakes** (right and left) to define your sectors of fire. Sector stakes prevent accidental firing into friendly positions. Tree limbs about 46 cm (18 in) long make good stakes. The stakes must be sturdy and must stick out of the ground high enough to keep your rifle from being pointed out of your sector.

Hammer in **aiming stakes** to help you fire into dangerous approaches at night and at other times when visibility is poor. Forked tree limbs about 30 cm (12 in) long make good stakes. Put one stake near the edge of the hole to rest the stock of your rifle on. Then put another stake forward of the rear (first) stake toward each dangerous approach. The forward stakes are used to hold the rifle barrel. To change the direction of your fire from one approach to another, move the rifle barrel from one forward stake to another. Leave the stock of the rifle on the rear stake.

AIMING AND SECTOR STAKES

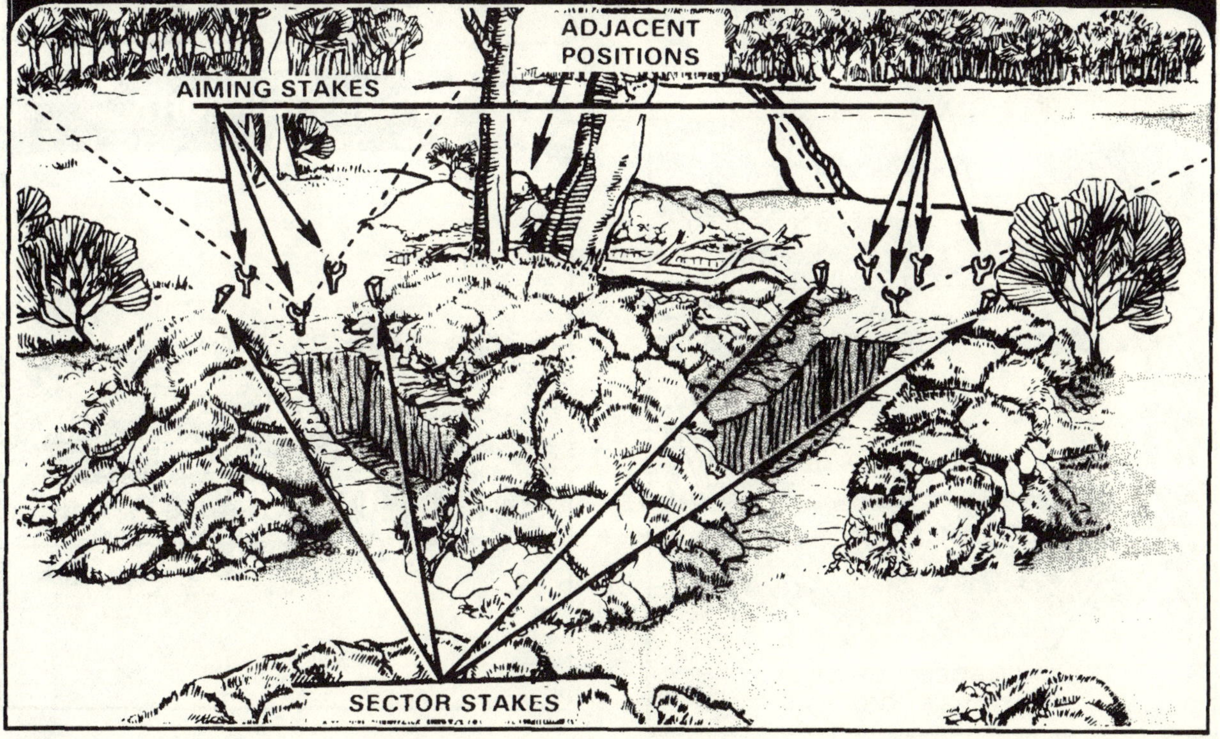

Dig two **grenade sumps** in the floor (one on each end). If the enemy throws a grenade into the hole, kick or throw it into one of the sumps. The sump will absorb most of the blast. The rest of the blast will be directed straight up and out of the hole. **Dig the grenade sumps:**

- **As wide as the entrenching tool blade.**
- **At least as deep as an entrenching tool.**
- **As long as the position floor is wide.**

For **water drainage,** slope the floor of the hole toward the grenade sumps. This may also cause grenades to roll into the sumps.

Build **overhead cover** for protection against airburst fragments. Build the overhead cover either across the center of the hole or off to its flanks.

When center overhead cover would not make a position easy to detect, build it. Put support logs 10 to 15 cm (4 to 6 in) in diameter on top of each other along the entire length of the frontal and rear cover.

Put logs 10 to 15 cm (4 to 6 in) in diameter side by side across the support logs as the base for the overhead cover.

BUILDING CENTER OVERHEAD COVER

Put a water-repellent layer, such as C-ration boxes or a poncho, over the base logs. This helps keep water from leaking through the overhead cover.

WATERPROOFING

Then put 15 to 20 cm (6 to 8 in) of dirt on top of the waterproofing material. Finally, mold and camouflage the cover to blend with the terrain.

When center overhead cover would make your position easy to detect, build flank overhead cover. That method gives both you and your buddy your own overhead cover. However, neither of you can observe or fire into your sectors of fire while under it.

When flank overhead cover is used, dig only one grenade sump. Dig it in the center of the floor against the back wall and slope the floor toward it.

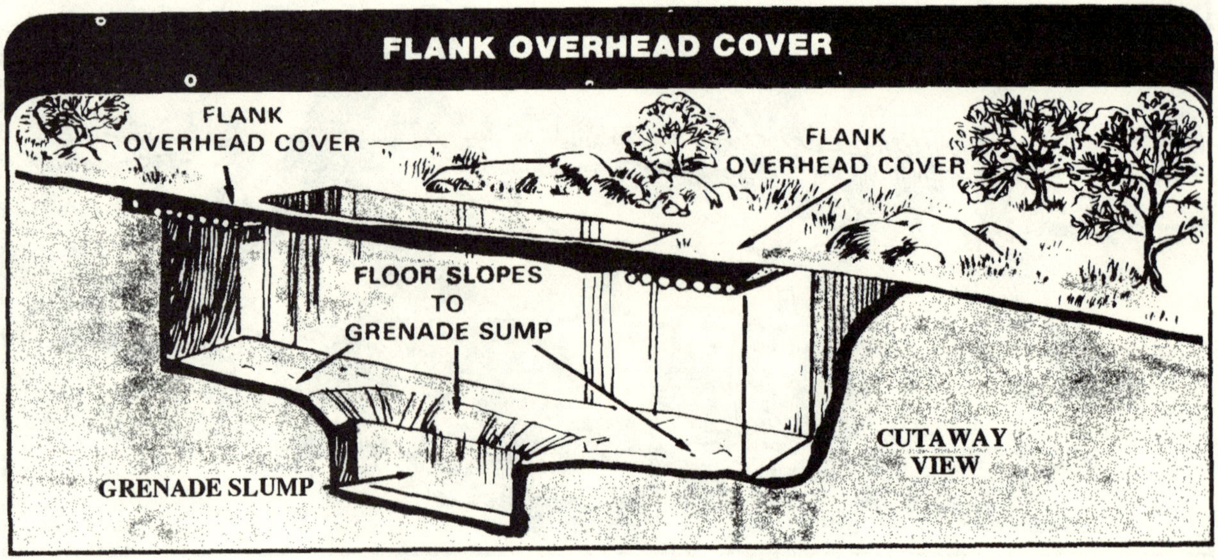

Dig out an area for flank overhead cover at each end of **the position:**

- **About 30 cm (12 in) deep.**
- **Long enough to extend about 45 cm (18 in) beyond both sides of the hole.**
- **About 1 meter (3 ft) wide.**

Save the sod for camouflage.

DIGGING OUT FOR FLANK OVERHEAD COVER

Next, place support logs, about 10 to 15 cm (4 to 6 in) in diameter, across the dug-out holes. This will support the rest of the overhead cover material. Put a water-repellent layer, such as C-ration boxes or a poncho, over the support logs. This helps keep water from leaking through the overhead cover.

SUPPORTING LOGS

Then put 15 to 20 cm (6 to 8 in) of dirt on top of the waterproofing material. Cover the dirt with the sod and camouflage it.

CAMOUFLAGING WITH SOD

Then get in the hole and dig out a cave-like compartment at each end of the position under the overhead cover. Dig your compartment large enough for you and your equipment. Dig your buddy's compartment large enough for him and his equipment.

COMPARTMENT

In sandy or loose soil, the sides of your position may require **revetments** to keep them from collapsing. Use such things as mesh wire, boards, or logs for revetting. Tie anchor string, rope, or wire to the revetting material and stake them down. Drive the stakes into the ground. This hides them and keeps them from being mistaken for aiming stakes or sector stakes.

REVETMENT

ONE-MAN FIGHTING POSITION

Sometimes you may have to build and occupy a one-man fighting position. Except for its size, a one-man position is built the same way as a two-man fighting position. The hole of a one-man position is only large enough for you and your equipment.

MACHINE GUN FIGHTING POSITION

If you are in a machine gun crew, you and the other members must build a machine gun fighting position. However, before you can start **work on the position, your leader must:**

- **Position the machine gun.**
- **Assign it a primary (and a secondary, if required) sector of fire.**
- **Assign it a principal direction of fire (PDF) or final protective line (FPL).**

NOTE: The FPL is a line on which the gun fires grazing fire across the unit's front. Grazing fire is fired 1 meter above the ground. When an FPL is not assigned, a PDF is. A PDF is a direction toward which the gun must be pointed when not firing at targets in other parts of its sector.

The first thing to do when building a machine gun position is to mark the position of the tripod legs. Then mark the sectors of fire with sector stakes, and trace the outline of the hole and its frontal cover on the ground.

TRACING OUTLINE

For an M60 machine gun position, dig two firing platforms for the gun. One platform is on the primary sector of fire side of the position, and the machine gun tripod is used on this platform. The other platform is on the secondary sector of fire side of the position, and the machine gun biped is used when firing on this platform. A trench must be dug for the bipod legs.

POSITION WITH FIRING PLATFORMS

The firing platforms reduce the profile of the gunner. They also reduce the height of the frontal cover needed. The firing platforms must not, however, be so low that the gun cannot be traversed across its sector of fire.

In some cases, the floor of the platforms may need to be lined with sandbags. Also, sandbags may be needed on each tripod leg to keep it from moving.

GRENADE SUMP LOCATIONS

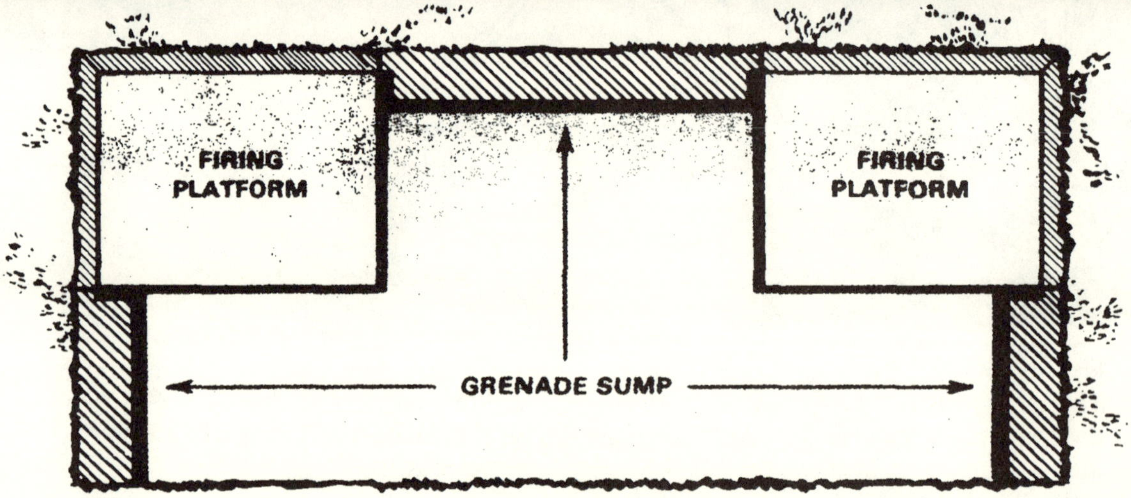

TOP VIEW

After the firing platforms have been dug, prepare your range card (app I) and then dig your hole. Dig the hole in the shape of an inverted T. The top of the T, however, must be longer than the shaft of the T. Dig the hole deep enough to, protect the crew and still let the gunner fire the machine gun (usually about armpit deep). Use the dirt from the hole to build frontal, flank, and rear cover. The frontal cover is built first. When the frontal cover is high and thick enough, use the rest of the dirt to build flank and rear cover.

Dig three grenade sumps, one at each end of the T. Dig the grenade sumps like those in a two-man fighting position.

Build the overhead cover for the position like that for a two-man fighting position.

MACHINE GUN POSITION WITH OVERHEAD COVER

When an M60 machine gun has only one sector of fire, dig only half of the position (only one firing platform).

NO SECONDARY SECTOR

When there is a three-man crew for an M60 machine gun, the third man (the ammunition bearer) digs a one-man fighting position. Usually, his position is on the same side of the machine gun as its FPL or PDF. From that position, he can observe and fire into the machine gun's secondary sector and, at the same time, see the gunner and assistant gunner. The ammunition bearer's position is connected to the machine gun position by a crawl trench so that he can bring ammunition to the gun or replace the gunner or the assistant gunner.

POSITION FOR AMMUNITION BEARER

In a caliber .50 machine gun position, dig only one firing platform for the gun. Dig the platform below ground level, like that for an M60 machine gun except deeper. Because of the gun's vibrations, you may have to line the floor of the platform with sandbags. Sandbags may also be needed on each tripod leg to keep it from moving. Also, the walls of the platform may need revetments.

After digging the platform, prepare your range card and then dig your hole. The hole should be the shape of an L, with the platform in the center of the L. Dig the hole deep enough to protect the crew and still let the gunner fire the machine gun (usually about armpit deep). Use the dirt from the hole to build frontal, flank, and rear cover. Build the frontal cover first. When that is completed, use the rest of the dirt to build flank and rear cover.

MODIFIED TWO-MAN FIGHTING POSITION

Dig two grenade sumps, one at both ends of the L, like those in a two-man fighting position.

Build the overhead cover like that for a two-man fighting position.

MODIFIED TWO-MAN FIGHTING POSITION WITH OVERHEAD COVER

The Dragon can be fired from either a **one-man** or a **two-man** fighting position. However, you must make some changes in the positions. Like the machine gun, a Dragon needs a range card. Prepare it before digging your hole.

Dig the hole wide enough to let the muzzle end of the launcher extend 15 cm (6 in) beyond the front of the hole and the rear of the launcher extend out over the rear of the hole. This is to keep the backblast out of the hole.

DRAGON POSITION

Dig the hole only **waist deep** on the side the Dragon will be fired from. This lets you move while tracking. Dig the other side of the hole armpit deep. Also, dig a small hole for the biped legs in front of the hole. Because of your height above the ground when firing the Dragon, build frontal cover high enough to hide you and, if feasible, the backblast.

Build **overhead cover** on the flanks of the position. Build it large enough for you, your equipment, and the Dragon. Overhead cover is not usually built across the center of the hole in a Dragon position. The center overhead cover would have to be so high that it would be easy for the enemy to spot.

FIRING DRAGON FROM POSITION

Clear the backblast area before firing the weapon. That means checking to see if any soldiers are in the backblast area or if any walls, large trees, or other things are in a position to deflect the backblast. If the weapon is to be fired from a two-man fighting position, make sure that the other soldier in the hole is not in the backblast area.

90-MM RECOILLESS RIFLE FIGHTING POSITION

Build a 90-mm recoilless rifle (RCLR) position like a Dragon position, but dig the hole a little longer when firing to the right side of the frontal cover. That lets the assistant gunner work from the right side of the RCLR. Prepare your range card before digging the hole. Also, clear the backblast area before firing the RCLR.

90-MM RECOILLESS RIFLE POSITION

LIGHT ANTITANK WEAPON (M72A2) AND FLAME ASSAULT SHOULDER WEAPON (FLASH) FIGHTING POSITION

There is no special fighting position for the M72A2 or FLASH. They can be fired from any fighting position. Before firing any of these weapons, clear the backblast area.

FIRING LAW FROM POSITION

TRENCHES

When there is time, dig trenches to connect fighting positions. Trenches provide covered routes between positions. The depth of the trenches depends on the time and type of help and equipment available to dig them. Without engineer help, crawl trenches about 1 meter (3 feet) deep and two thirds of a meter (2 feet) wide are probably all that can be dug. Dig the trenches zigzagged so that the enemy will not be able to fire down a long section if he gets into the trench, and so that shrapnel from shell bursts will lose some of its effectiveness.

AIR VIEW OF TRENCHES

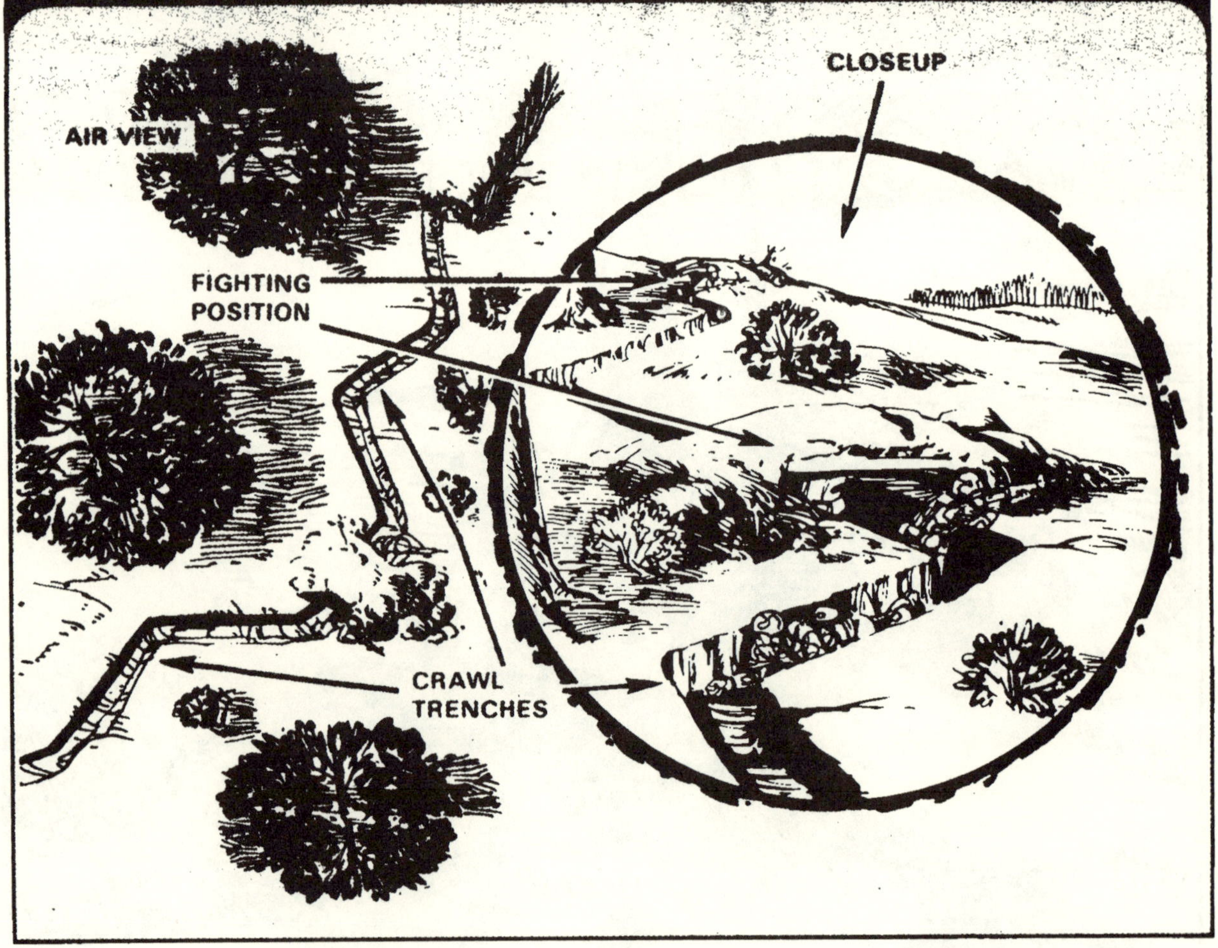

STORAGE COMPARTMENTS

A fighting position should have a place for storing equipment and ammunition. When your position has overhead cover across its center, dig a storage compartment in the bottom of the back wall. The size of the compartment depends on the amount of equipment and ammunition to be stored.

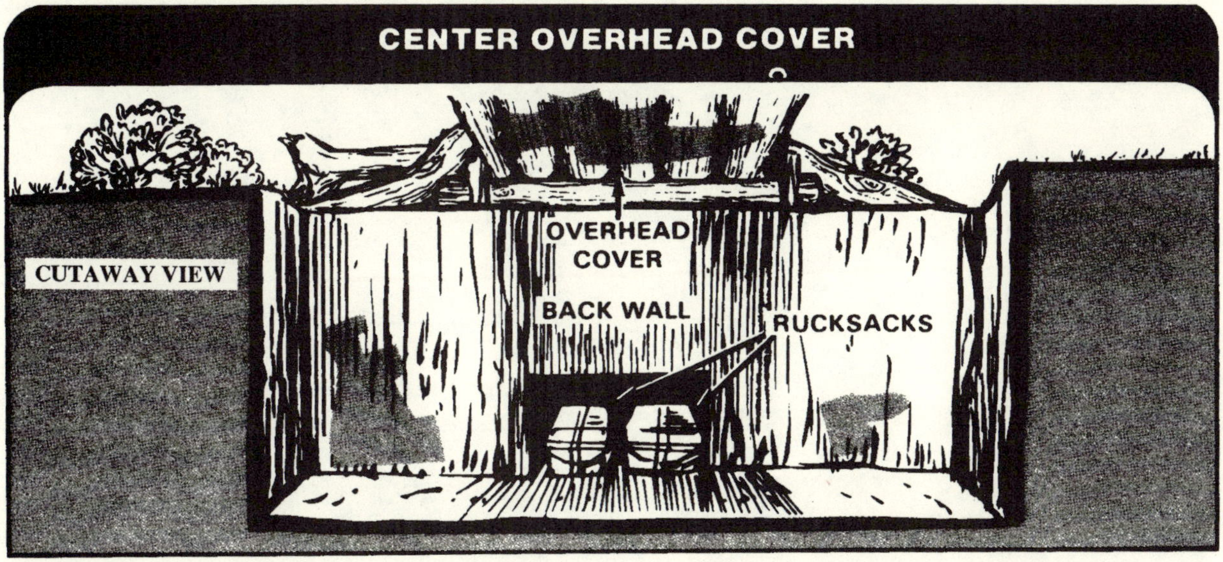

When your position has flanked overhead cover, use the compartments dug for the overhead cover as storage compartments.

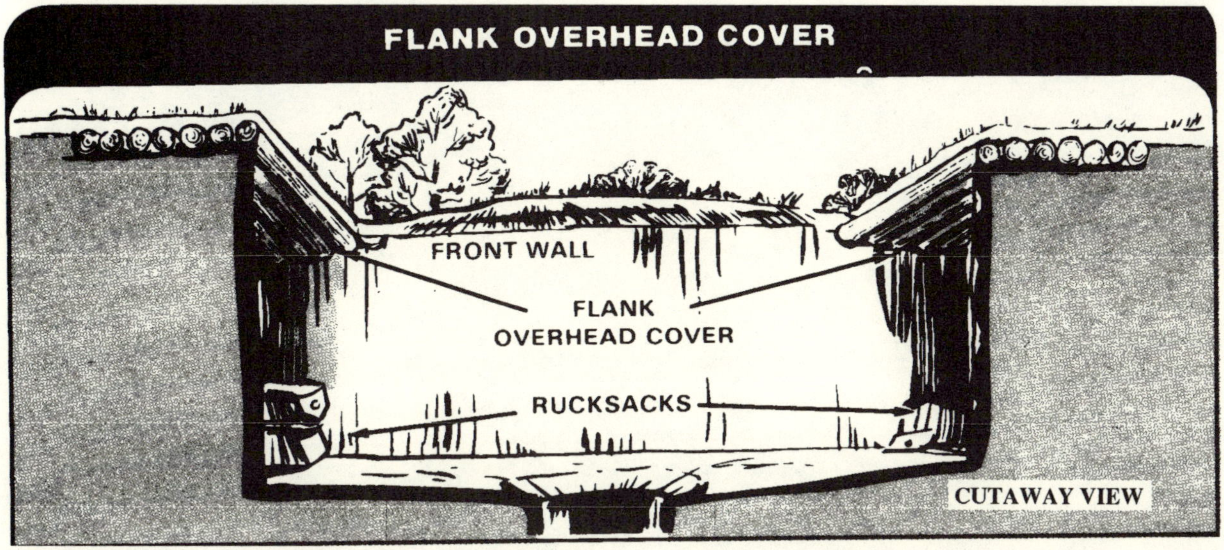

If you dig your storage compartment large enough, it may provide extra space in which you can stretch out while sleeping. This lets you sleep inside the position and under cover.

CHAPTER 3

Movement

GENERAL

Normally, you will spend more time moving than fighting. You must use proper movement techniques to avoid contact with the enemy when you are not prepared for contact.

The fundamentals of movement discussed in this chapter provide techniques that all soldiers should learn. These techniques should be practiced until they become second nature.

CONTENTS

GENERAL 3-1
MOVEMENT TECHNIQUES 3-2
IMMEDIATE ACTIONS WHILE MOVING 3-5
MOVING WITHIN A TEAM 3-8
FIRE AND MOVEMENT 3-9
MOVING WITH TANKS 3-10

MOVEMENT TECHNIQUES

Your unit's ability to move depends on your movement skills and those of your fellow soldiers. Use the following techniques to avoid being seen or heard **by the enemy:**

- **Camouflage yourself and your equipment.**
- **Tape your dog tags together and to the chain so they cannot slide or rattle. Tape or pad the parts of your weapon and equipment that rattle or are so loose that they may snag (the tape or padding must not interfere with the operation of the weapon or equipment). Jump up and down and listen for rattles.**
- **Wear soft, well-fitting clothes.**
- **Do not carry unnecessary equipment. Move from covered position to revered position (taking no longer than 3 to 5 seconds between positions).**
- **Stop, look, and listen before moving. Look for your next position before leaving a position.**
- **Look for covered and concealed routes on which to move.**
- **Change direction slightly from time to time when moving through tall grass.**
- **Stop, look, and listen when birds or animals are alarmed (the enemy may be nearby).**
- **Use battlefield noises, such as weapon noises, to conceal movement noises.**
- **Cross roads and trails at places that have the most cover and concealment (large culverts, low spots, curves, or bridges).**
- **Avoid steep slopes and places with loose dirt or stones.**
- **Avoid cleared, open areas and tops of hills and ridges.**

METHODS OF MOVEMENT

In addition to walking, you may move in one of three other methods — low crawl, high crawl, or rush.

The **low crawl** gives you the lowest silhouette. Use it to cross places where the concealment is very low and enemy fire or observation prevents you from getting up. Keep your body flat against the ground. With your firing hand, grasp your weapon sling at the upper sling swivel. Let the front handguard rest on your forearm (keeping the muzzle off the ground), and let the weapon butt drag on the ground.

To move, push your arms forward and pull your firing side leg forward. Then pull with your arms and push with your leg. Continue this throughout the move.

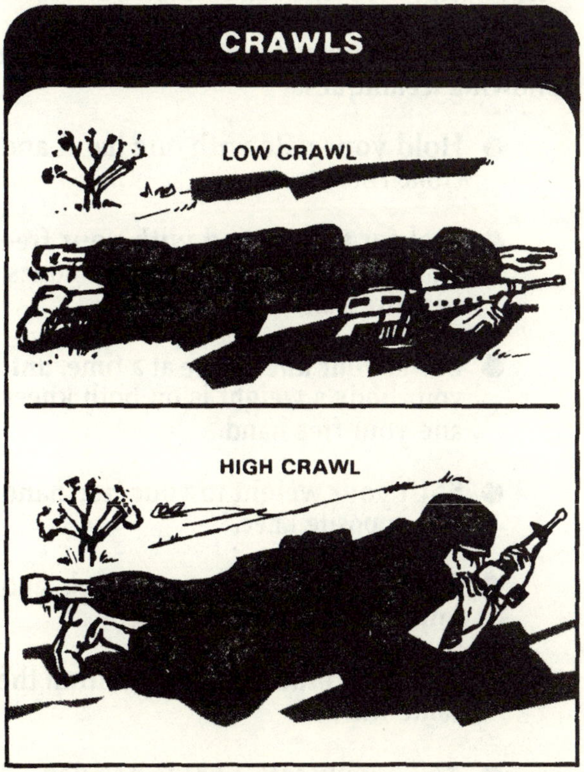

The **high crawl** lets you move faster than the low crawl and still gives you a low silhouette. Use this crawl when there is good concealment but enemy fire prevents you from getting up. Keep your body off the ground and resting on your forearms and lower legs. Cradle your weapon in your arms and keep its muzzle off the ground. Keep your knees well behind your buttocks so your body will stay low.

To move, alternately advance your right elbow and left knee, then your left elbow and right knee.

The **rush** is the fastest way to move from one position to another. Each rush should last from 3 to 5 seconds. The rushes are kept short to keep enemy machine gunners or riflemen from tracking you. However, do not stop and hit the ground in the open just because 5 seconds have passed. Always try to hit the ground behind some cover. Before moving, pick out your next covered and concealed position and the best route to it.

Make your move from the **prone position as follows:**

- **Slowly raise your head and pick your next position and the route to it.**
- **Slowly lower your head.**
- **Draw your arms into your body (keeping your elbows in).**
- **Pull your right leg forward.**
- **Raise your body by straightening your arms.**
- **Get up quickly.**
- **Run to the next position.**

When you are ready to stop moving, **do the following:**

- **Plant both of your feet.**
- **Drop to your knees (at the same time slide a hand to the butt of your rifle).**
- **Fall forward, breaking the fall with the butt of the rifle.**
- **Go to a prone firing position.**

If you have been firing from one position for some time, the enemy may have spotted you and may be waiting for you to come up from behind cover. So, before rushing forward, roll or crawl a short distance from your position. By coming up from another spot, you may fool an enemy who is aiming at one spot, waiting for you to rise.

When the route to your next position is through an open area, rush by zigzagging. If necessary, hit the ground, roll right or left, then rush again.

MOVING WITH STEALTH

Moving with stealth means moving quietly, slowly, and carefully. This requires great patience.

To move with stealth, use the **following techniques:**

- **Hold your rifle at port arms (ready position).**
- **Make your footing sure and solid by keeping your body's weight on the foot on the ground while stepping.**
- **Raise the moving leg high to clear brush or grass.**
- **Gently let the moving foot down toe first, with your body's weight on the rear leg.**
- **Lower the heel of the moving foot after the toe is in a solid place.**
- **Shift your body's weight and balance to the forward foot before moving the rear foot.**
- **Take short steps to help maintain balance.**

At night, and when moving through dense vegetation, avoid making noise. Hold your weapon with one hand, and keep the other hand forward, feeling for obstructions.

When going into a prone position, use the **following techniques:**

- **Hold your rifle with one hand and crouch slowly.**
- **Feel for the ground with your free hand to make sure it is clear of mines, tripwires, and other hazards.**
- **Lower your knees, one at a time, until your body's weight is on both knees and your free hand.**
- **Shift your weight to your free hand and opposite knee.**
- **Raise your free leg up and back, and lower it gently to that side.**
- **Move the other leg into position the same way.**
- **Roll quietly into a prone position.**

Use the following techniques **when crawling:**

- **Crawl on your hands and knees. Hold your rifle in your firing hand. Use your nonfiring hand to feel for and make clear spots for your hands and knees to move to.**

- **Move your hands and knees to those spots, and put them down softly.**

IMMEDIATE ACTIONS WHILE MOVING

This section furnishes guidance for the immediate actions you should take when reacting to enemy indirect fire and flares.

REACTING TO INDIRECT FIRE

If you come under indirect fire while moving, quickly look to your leader for orders. He will either tell you to run out of the impact area in a certain direction or will tell you to follow him. If you cannot see your leader, but can see other team members, follow them. If alone, or if you cannot see your leader or the other team members, run out of the area in a direction away from the incoming fire.

FOLLOWING A TEAM LEADER OUT OF IMPACT AREA

It is hard to move quickly on rough terrain, but the terrain may provide good cover. In such terrain, it may be best to take cover and wait for flares to burn out. After they burn out, move out of the area quickly.

REACTING TO GROUND FLARES

The enemy puts out ground flares as warning devices. He sets them off himself or attaches tripwires to them for you to trip on and set them off. He usually puts the flares in places he can watch.

If you are caught in the light of a ground flare, move quickly out of the lighted area. The enemy will know where the ground flare is and will be ready to fire into that area. Move well away from the lighted area. While moving out of the area, look for other team members. Try to follow or join them to keep the team together.

REACTING TO AERIAL FLARES

The enemy uses aerial flares to light up vital areas. They can be set off like ground flares; fired from hand projectors, grenade launchers, mortars, and artillery; or dropped from aircraft.

If you hear the firing of an aerial flare while you are moving, hit the ground (behind cover if possible) while the flare is rising and before it bursts and illuminates.

If moving where it is easy to blend with the background (such as in a forest) and you are caught in the light of an aerial flare, freeze in place until the flare burns out.

If you are caught in the light of an aerial flare while moving in an open area, immediately crouch low or lie down.

If you are crossing an obstacle, such as a barbed-wire fence or a wall, and get caught in the light of an aerial flare, crouch low and stay down until the flare burns out.

The sudden light of a bursting flare may temporarily blind both you and the enemy. When the enemy uses a flare to spot you, he spoils his own night vision. To protect your night vision, close one eye while the flare is burning. When the flare burns out, the eye that was closed will still have its night vision.

REACTING TO AERIAL FLARES

MOVING WITHIN A TEAM

You will usually move as a member of a team. Small teams, such as infantry fire teams, normally move in a wedge formation. Each soldier in the team has a set position in the wedge, determined by the type weapon he carries. That position, however, may be changed by the team leader to meet the situation. The normal distance between soldiers is 10 meters.

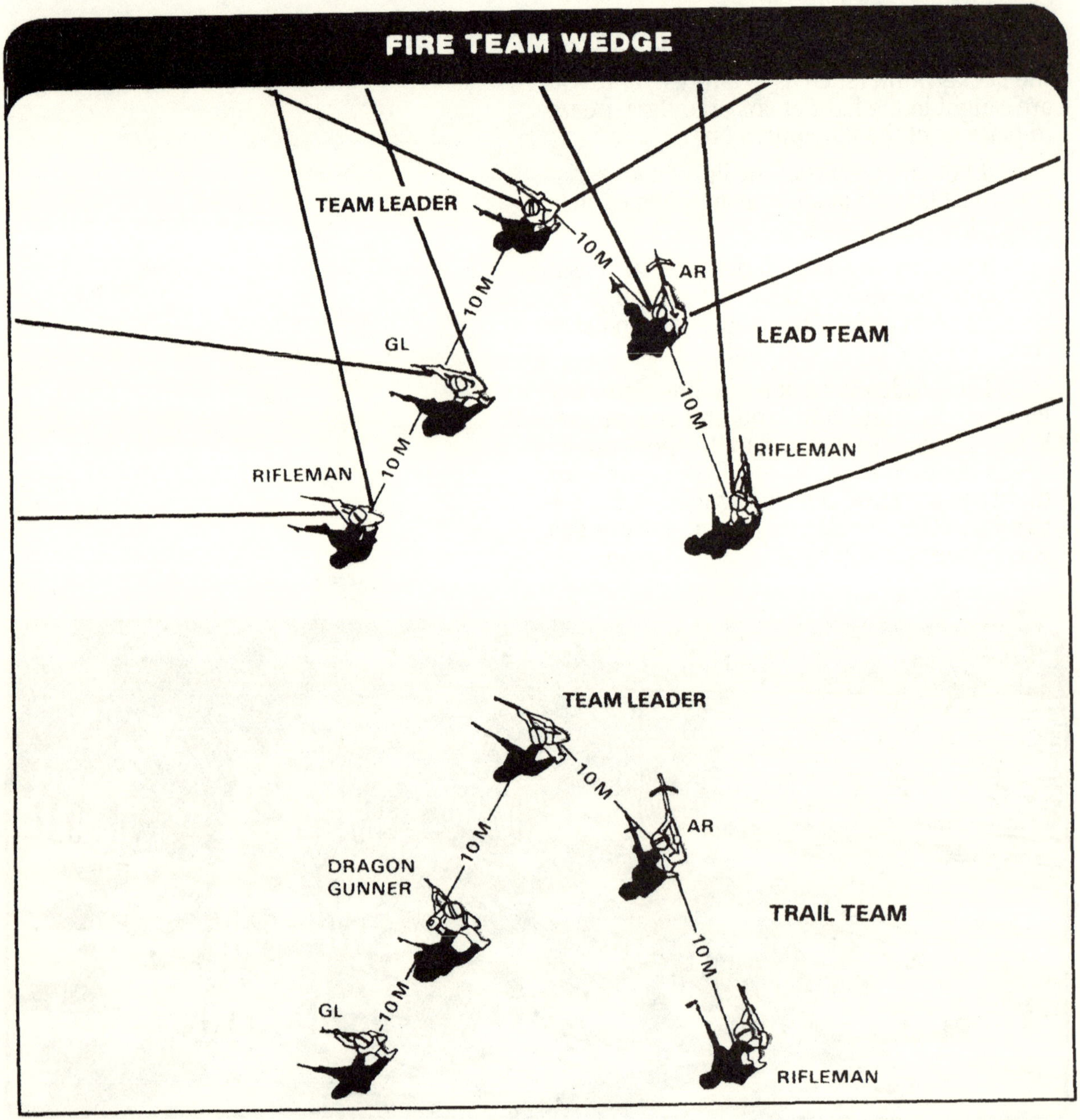

You may have to make a temporary change in the wedge formation when moving through close terrain. The soldiers in the sides of the wedge close into a single file when moving in thick brush or through a narrow pass. After passing through such an area, they should spread out, again forming the wedge. You should not wait for orders to change the formation or the interval. You should change automatically and stay in visual contact with the other team members and the team leader.

The team leader leads by setting the example. His standing order is, FOLLOW ME AND DO AS I DO. When he moves to the left, you should move to the left. When he gets down, you should get down. When he fires, you should fire.

When visibility is limited, control during movement may become difficult. Two l-inch horizontal strips of luminous tape, sewn directly on the rear of the helmet camouflage band with a l-inch space between them, are a device for night identification.

LUMINOUS NIGHT IDENTIFICATION DEVICE

Night identification for your patrol cap could be two l-inch by 1/2-inch strips of luminous tape sewn vertically, directly on the rear of the cap. They should be **centered,** with the bottom edge of each tape even with the bottom edge of the cap and with a l-inch space between the two tapes.

FIRE AND MOVEMENT

When a unit makes contact with the enemy, it normally starts firing at and moving toward the enemy. Sometimes the unit may move away from the enemy. That technique is called **fire and movement.** It is conducted either to close with and destroy the enemy, or to move away from the enemy so as to break contact with him.

The firing and moving take place at the same time. There is a fire element and a movement element. These elements may be single soldiers, buddy teams, fire teams, or squads. Regardless of the size of the elements, the action is still fire and movement.

The **fire element** covers the move of the movement element by firing at the enemy. This helps keep the enemy from firing back at the movement element.

The **movement element** moves either to close with the enemy or to reach a better position from which to fire at him. The movement element should not move until the fire element is firing.

Depending on the distance to the enemy position and on the available cover, the fire element and the movement element switch roles as needed to keep moving.

Before the movement element moves beyond the supporting range of the fire element (the distance within which the weapons of the fire element can fire and support the movement element), it should take a position from which it can fire at the enemy. The movement element then becomes the next fire element and the fire element becomes the next movement element.

If your team makes contact, your team leader should tell you to fire or to move. He should also tell you where to fire from, what to fire at, or where to move to. When moving, use the low crawl, high crawl, or rush.

MOVING WITH TANKS

You will often have to move with tanks. When you must move as fast as the tanks, you should ride on them. However, riding on a tank makes you vulnerable to all types of fire. It also reduces the tank's maneuverability and the ability to traverse its turret. If contact is made with the enemy, you must dismount from the tank at once.

To mount a tank, first get permission from the tank commander. Then mount from the tank's right front, not its left side where the coax machinegun is mounted. Once mounted, move to the rear deck, stand, and hold on to the bustle rack. If there is not enough room for you on the rear deck, you may have to stand beside the turret and hold onto a hatch or hatch opening.

When riding on a tank, be alert for trees that may knock you off and obstacles that may cause the tank to turn suddenly. Also be alert for enemy troops that may cause the tank to travers its turret quickly and fire.

Riding on a tank is always hazardous and should be done only when the risks of riding are outweighed by the advantages of riding.

TROOPS RIDING TANKS

CHAPTER 4

Observation

GENERAL

During all types of operations, you will be looking for the enemy. However, there will be times when you will be posted in an observation post (OP) to watch for enemy activity.

An OP is a position from which you watch an assigned sector of observation and report all activity seen or heard in your sector. Chapter 6 provides guidance on collecting and reporting information learned by observation.

CONTENTS

GENERAL 4-1
HOW TO OBSERVE 4-2
THINGS TO LOOK AND LISTEN FOR 4-4
RANGE ESTIMATION 4-4

HOW TO OBSERVE

This section discusses the techniques you will use for day and night observation.

DAY OBSERVATION

In daylight, use the **visual search technique** to search terrain. **Do this in two steps:**

- **Step 1.** Make a quick, overall search of the entire sector for obvious targets and unnatural colors, outlines, or movements. Look first at the area just in front of your position, and then quickly scan the entire area out to the maximum range you want to observe. If the sector is wide, divide it and search each subsector as in **Step 2.**

OVERALL SEARCH

- **Step 2.** Observe overlapping, 50-meter-wide strips, alternating from left to right and right to left, until you have searched the entire sector. When you see a suspicious spot, search it well.

OVERLAPPING 50-METER SEARCH

NIGHT OBSERVATION

At night, use anyone of three night observation techniques to search terrain.

Dark Adaptation Technique. First, let your eyes become adjusted to the darkness. Do so by staying either in a dark area for about 30 minutes, or in a red-lighted area for about 20 minutes followed by about 10 minutes in a dark area. The red-lighted method may save time by allowing you to get orders, check equipment, or do some other job before moving into darkness.

Off-Center Vision Technique. Focus your attention on an object but look slightly away from it. The object will be more visible this way than when you look straight at it.

Scanning Technique. Again focus your attention on an object, but do not look directly at it. Now move your eyes in short, abrupt, and irregular movements around it, pausing a few seconds after each move.

THINGS TO LOOK AND LISTEN FOR

In trying to find the enemy in a sector of observation, look and listen for **these signs of his presence:**

- **Sounds.**
- **Dust or vehicle exhaust.**
- **Movement.**
- **Positions.**
- **Outlines or shadows.**
- **Shine or glare.**
- **Contrasting colors.**

SOUNDS

Listen for such things as footsteps, limbs or sticks breaking, leaves rustling, men coughing, and equipment or vehicle sounds. These may be hard to distinguish from other battlefield and animal sounds.

Sounds can alert you to the direction or general location of the enemy. They may not pinpoint his exact location. However, if a sound alerts you, you are more apt to spot the enemy.

DUST OR VEHICLE EXHAUST

Moving foot soldiers or vehicles often raise dust. Vehicle exhaust smoke also rises. You can spot dust and vehicle smoke at long ranges.

MOVEMENT

Look for movement in your sector. Use the visual search technique.

POSITIONS

Look for enemy positions in obvious places, such as road junctions, hilltops, and lone buildings. Also look at areas with cover and concealment, such as woods and draws.

OUTLINES OR SHADOWS

Look for outlines or shadows of enemy soldiers, equipment, vehicles, or guns. The enemy may use the shadows of trees or buildings to hide himself and his equipment. Look for him in shaded areas.

SHINE OR GLARE

In darkness, look for light sources such as burning cigarettes, headlights, or flashlights. In daylight, look for reflected light or glare from smooth, polished surfaces such as windshields, headlights, mess gear, watch crystals, or uncamouflaged skin.

CONTRASTING COLORS

Look for contrasts between background color and the colors of uniforms, equipment, and skin. For example, a soldier's T-shirt or towel may contrast with its background.

RANGE ESTIMATION

You must often estimate ranges. Your estimates will be easier to make and more accurate if you use the 100-meter unit-of-measure method, the appearance-of-objects method, or the flash-and-sound method. This section discusses the use of these methods.

1OO-METER UNIT-OF-MEASURE METHOD (DAYTIME)

Picture a distance of 100 meters on the ground. For ranges up to 500 meters, count the number of 100-meter lengths between the two points you want to measure. Beyond 500 meters, pick a point halfway to the target, count the number of 100-meter lengths to the halfway point, and then double that number to get the range to the target.

Sloping ground changes the appearance of 100-meter lengths. Ground that slopes upward makes them look longer than 100 meters, and ground that slopes downward makes them look shorter than 100 meters. Thus, the tendency is to underestimate 100-meter lengths on upslopes and overestimate them on downslopes.

The accuracy of the 100-meter method depends on how much ground is visible. This is most true at long ranges. If a target is at a range of 500 meters or more, and you can only see part of the ground between yourself and the target, it is hard to use this method with accuracy.

APPEARANCE-OF-OBJECTS METHOD (DAYTIME)

This method is a way to estimate range by the apparent size and detail of an object. It is a common method that is used in everyday life. For example, a motorist trying to pass another car judges the distance of oncoming cars based on their apparent size. He is not interested in exact distances, but only in having enough room to safely pass the car in front of him. Suppose he knows that at a distance of 1 mile an oncoming car appears to be 1 inch wide and 2 inches high, with a half inch between the headlights. Then, any time he sees an oncoming car that fits those dimensions, he knows it is about 1 mile away.

The same technique can be used to estimate ranges on the battlefield. If you know the apparent size and detail of troops and equipment at known ranges, then you can compare those characteristics to similar objects at unknown ranges. When the characteristics match, the range does also.

To use the appearance-of-objects method, you must be familiar with characteristic details of objects as they appear at various ranges. As you must be able to see those details to make the method work, anything that limits visibility (such as weather, smoke, or darkness) will limit the effectiveness of this method.

COMBINATION OF METHODS

Battlefield conditions are not always ideal for estimating ranges. If the terrain limits the use of the 100-meter unit-of-measure method, and poor visibility limits the use of the appearance-of-objects method, you may have to use a combination of methods. For example, if you cannot see all of the terrain out to the target, you can still estimate distance from the apparent size and detail of the target itself. A haze may obscure the target details, but you may still be able to judge its size or use the 100-meter method. By using either one or both of the methods, you should arrive at a figure close to the true range.

FLASH-AND-SOUND METHOD (BEST AT NIGHT)

Sound travels through air at 300 meters (1,100 feet) per second. That makes it possible to estimate distance if you can both see and hear a sound-producing action.

When you see the flash or smoke of a weapon, or the dust it raises, immediately start counting. Stop counting when you hear the sound associated with the action seen. The number at which you stop should be multiplied by three. This gives you the approximate distance to the weapon in hundreds of meters. If you stop at one, the distance is about 300 meters. If you stop at three, the distance is about 900 meters. When you must count higher than nine, start over again after counting nine (counting higher numbers throws the timing off).

CHAPTER 5

Nuclear, Biological, And Chemical Warfare

GENERAL

Nuclear, biological, and chemical (NBC) weapons can cause casualties, destroy or disable equipment, restrict the use of terrain, and disrupt operations. You must be prepared to fight and survive in an environment where NBC weapons have been used.

This chapter prescribes active and passive protection measures that will avoid or reduce the effects of NBC weapons.

CONTENTS

GENERAL 5-1
NUCLEAR WEAPONS 5-2
CHEMICAL AND BIOLOGICAL WEAPONS 5-5

NUCLEAR WEAPONS

This section describes the characteristics of nuclear explosions and their effects on soldiers, equipment, and supplies, and gives hasty measures for protection against nuclear attacks.

CHARACTERISTICS OF NUCLEAR EXPLOSIONS

The four main characteristics of **nuclear explosions are:**

- **BLAST (an intense shock wave).**
- **THERMAL RADIATION (heat and light).**
- **NUCLEAR RADIATION (radioactive material).**
- **EMP (electrical power surge).**

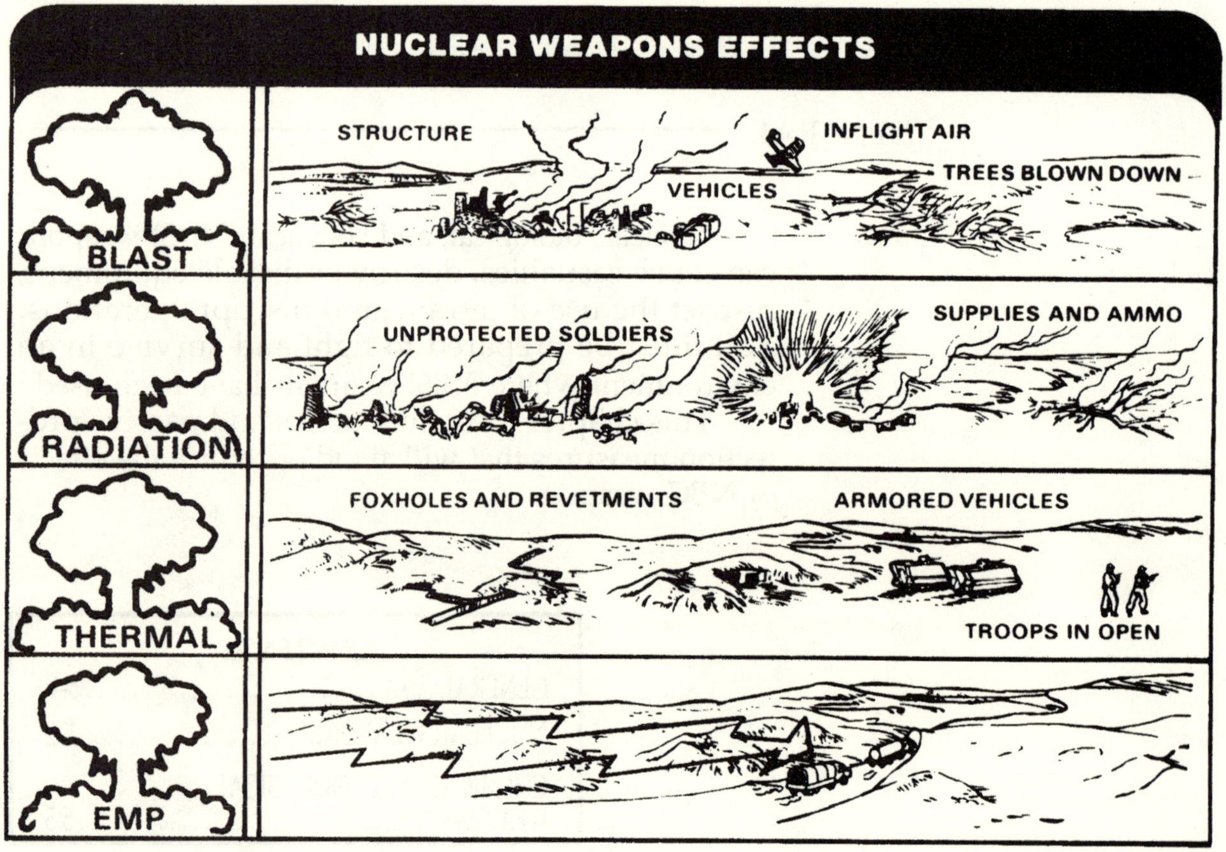

Blast produces an intense shockwave and high winds that create flying debris. It may collapse shelters and some fighting positions.

Thermal radiation causes burns and starts fires. The bright flash at the time of the explosion can cause a temporary loss of vision or permanent eye damage if you look at the explosion, especially at night.

Nuclear radiation can cause casualties and delay movements. It may last for days and cover large areas of terrain. It occurs in two stages: **initial and residual.**

- **Initial radiation is emitted directly from the fireball in the first minute after the explosion. It travels at the speed of light along straight lines and has high penetrating power.**

- **Residual radiation lingers after the first minute. It comes from the radioactive material originally in a nuclear weapon or from material, such as soil and equipment, made radioactive by the nuclear explosion.**

EMP is a massive surge of electrical power. It is created the instant a nuclear detonation occurs and is transmitted at the speed of light in all directions. It can damage solid-state components of electrical equipment (radios, radars, computers, vehicles) and weapon systems (TOW and Dragon). Equipment can be protected by disconnecting it from its power source and placing it in or behind some type of shielding material (armored vehicle or dirt wall) out of the line of sight to the burst. If no warning is received prior to a detonation, there is no effective means of protecting operating equipment.

EFFECTS ON SOLDIERS

The exposure of the human body to nuclear radiation causes damage to the cells in all parts of the body. This damage is the cause of "radiation sickness." The severity of this sickness depends on the radiation dose received, the length of exposure, and the condition of the body at the time. The early symptoms of radiation sickness will usually appear 1 to 6 hours after exposure. Those symptoms may include headache, nausea, vomiting, and diarrhea. Early symptoms may then be followed by a latent period in which the symptoms disappear. There is no first aid for you once you have been exposed to nuclear radiation. The only help is to get as comfortable as possible while undergoing the early symptoms.

If the radiation dose was small, the symptoms, if any, will probably go away and not recur. If the symptoms recur after a latent period, you should go to an aid station.

EFFECTS ON EQUIPMENT AND SUPPLIES

Blast can crush sealed or partly sealed objects like food cans, barrels, fuel tanks, and helicopters. Rubble from buildings being knocked down can bury supplies and equipment.

Heat can ignite dry wood, fuel, tarpaulins, and other flammable material. **Light** can damage eyesight.

Nuclear radiation can contaminate food and water.

PROTECTION AGAINST NUCLEAR ATTACKS

The best hasty protection against a nuclear attack is to take cover behind a hill or in a fighting position, culvert, or ditch. If caught in the open, drop flat on the ground at once and close your eyes. Cover exposed skin and keep your weapon under your body to avoid loss. If you know the direction of the burst, drop with your head away from the burst. Stay down until the blast wave passes, then check for injuries and equipment damage and prepare to continue the mission. See chapter 2 for additional considerations in the building of your fighting positions.

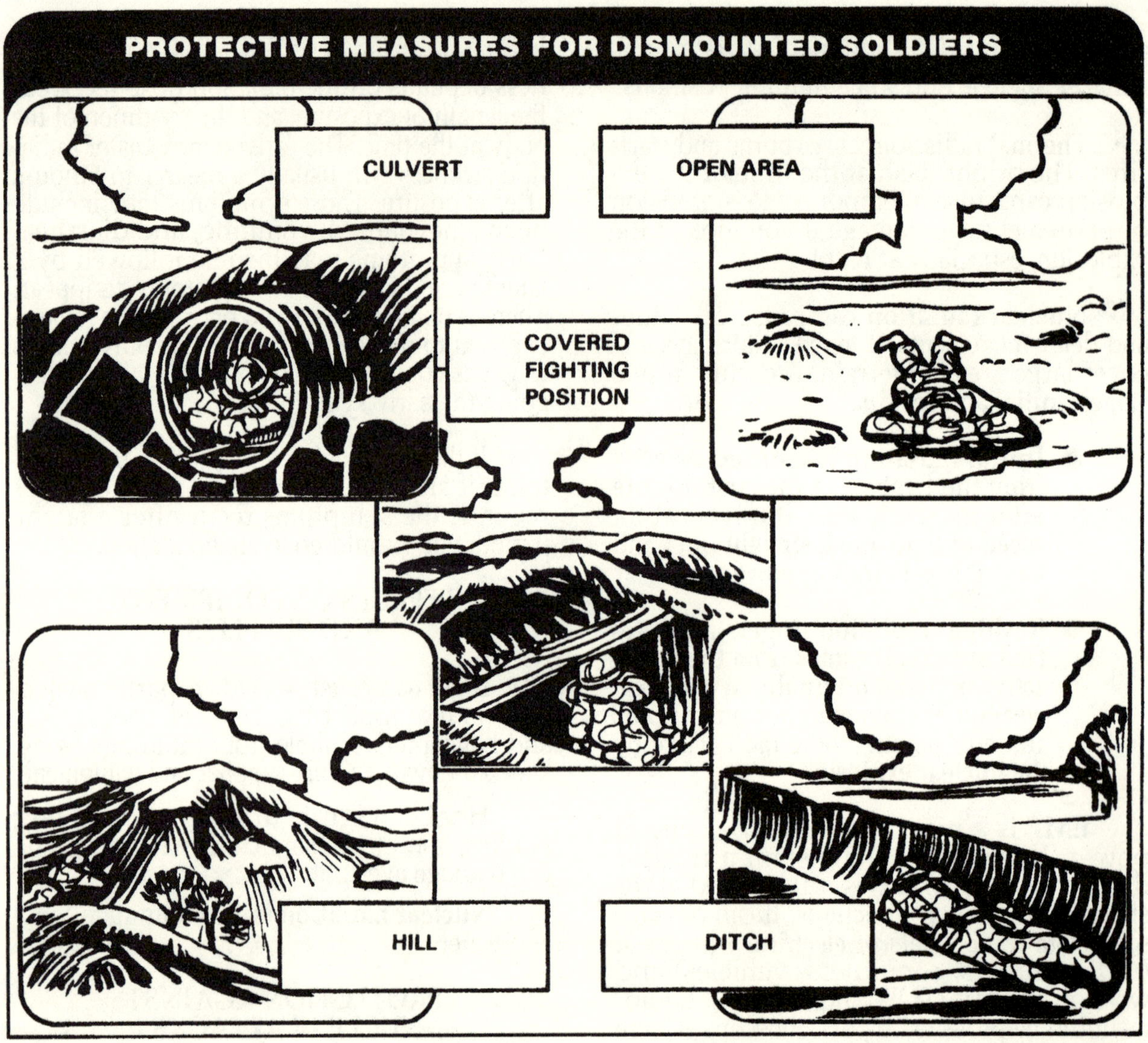

Radiation is the only direct nuclear effect that lingers after the explosion. As it cannot be detected by the senses, use radiac equipment to detect its presence. Procedures for detection can be found in FM 3-12 and FM 21-40. When feasible, move out of the contaminated area.

If your unit must stay in the contaminated area, it is best to stay in a dug-in position with overhead cover. When time does not permit constructing a well-prepared overhead cover, use a poncho. Stay under cover. When the fallout has finished falling, brush contamination off yourself and your equipment. Wash yourself and your equipment as soon as the mission permits.

CHEMICAL AND BIOLOGICAL WEAPONS

Enemy forces have both chemical and biological weapons. These weapons may be used separately or together, with or without nuclear weapons. Regardless of how they are used, you must be able to survive their effects and continue your mission.

CHARACTERISTICS OF CHEMICAL AND BIOLOGICAL AGENTS AND TOXINS

Chemical agents are like poisonous pesticides, but are far more powerful. They are meant to kill or injure you and are released to cover large areas. They may be released as gases, liquids, or sprays. The enemy may use a mixture of agents to cause confusion and casualties. Artillery, mortars, rockets, missiles, aircraft, bombs, and land mines can deliver the agents.

Biological agents are disease-producing germs. They create a disease hazard where none exists naturally. They may be dispersed as sprays by generators, or delivered by explosives, bomblets, missiles, and aircraft. They may also be spread by the release of germ-carrying flies, mosquitoes, fleas, and ticks. The US Army does not employ these agents, but other armies may.

Toxins are poisonous substances produced by living things (such as snake venom). Toxins are not living things and in this sense are chemicals. They would be used in combat in the same way as chemical-warfare agents, and they may disable or kill without warning.

Chemical and biological agents have little direct effect on equipment. Liquid chemical agents on your equipment can restrict its use until it is decontaminated.

EFFECTS ON TERRAIN

Liquid chemical agents may restrict the use of terrain and buildings.

It is difficult to decontaminate terrain. When time permits, it is best to wait for weather to decontaminate terrain naturally. Contaminated areas should be either bypassed or, when protective equipment is worn, crossed. After crossing a contaminated area, decontaminate yourself and your equipment as soon as the situation permits.

EFFECTS ON SOLDIERS

Chemical and biological agents may enter your body through your eyes, nose, mouth, or skin. They can disable or kill.

Liquid agents may be dispersed on you, your equipment, the terrain, and foliage. The agents may linger for hours or days and endanger you when you are unprotected.

Biological agents are hard to detect in early stages of use. If you find out or suspect that the enemy is using biological agents, report it to your leader.

The M8 automatic chemical-agent alarm can detect the presence of chemical agents in the air and produce an audible or visual signal. It will detect nerve, blood, and choking agents. The M43A1 detects only nerve-agent vapor. The use and maintenance of the M8 alarm is the responsibility of the unit NBC defense team.

DETECTION OF CHEMICAL AND BIOLOGICAL AGENTS

Your senses may not detect chemical agents, as most agents are odorless, colorless, tasteless, and invisible in battlefield concentrations. However, you can detect chemical agents by using the chemical-agent alarms and detection kits found in each company (FM 21-40).

CHEMICAL-AGENT ALARM

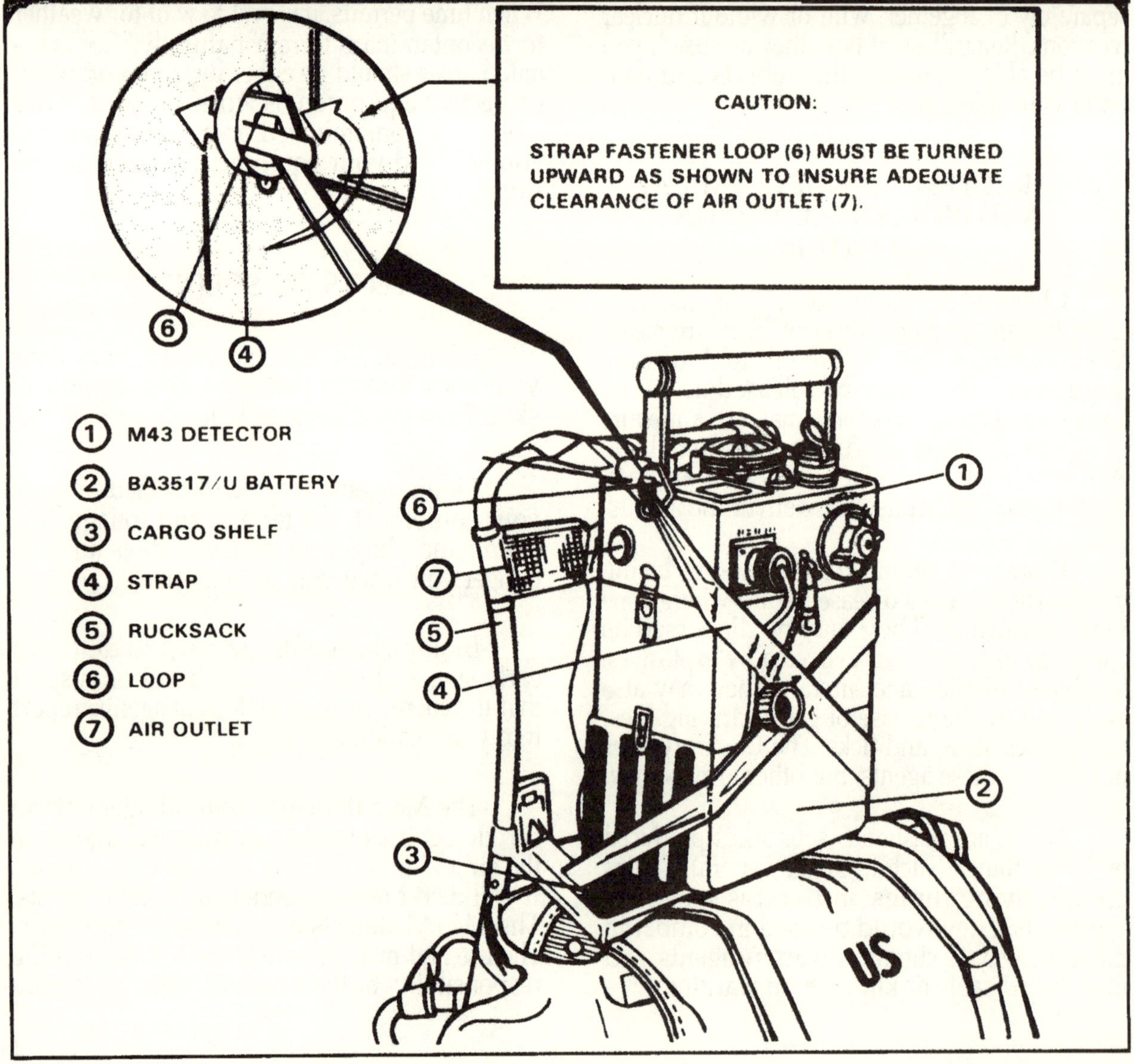

The ABC-M8 chemical-agent detector paper comes in a booklet of 25 sheets. It is a part of the M256 chemical-agent detector kit. The paper sheets turn dark green, yellow, or red on contact with liquid V-type nerve agents, G-type nerve agents, or blister (mustard) agents, respectively they do not detect vapor, The test is not always reliable on porous material such as wood or rubber. Many substances (including some solvents and decontaminants) can cause a color change in the paper, so such a change indicates only that a chemical agent may be present. Positive detector-paper tests should be verified by testing with chemical-agent detector kits.

CHEMICAL-AGENT DETECTOR PAPER

The M256 chemical-agent detector kit is issued to squads. It is used to detect dangerous vapor concentrations of nerve, blister, or blood agents. It should be used when the platoon or company is under chemical attack, when a chemical attack is reported to be likely, or when the presence of a chemical agent is suspected.

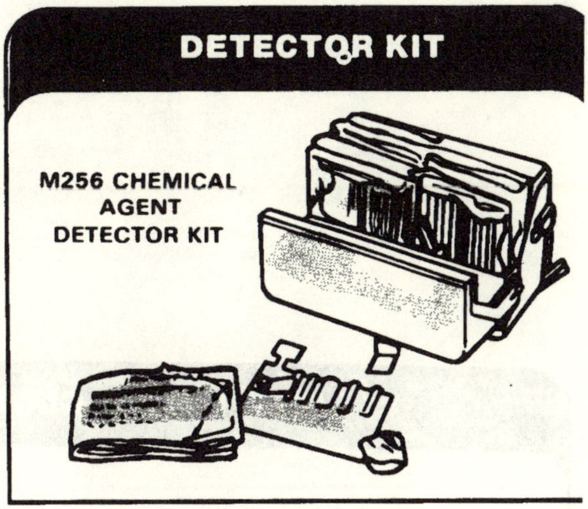

DETECTOR KIT

ALARMS

If you recognize or suspect a chemical or biological attack, STOP BREATHING, PUT YOUR MASK ON, CLEAR AND CHECK IT, AND GIVE THE ALARM set by your unit's SOP.

CB ALARM

PROTECTION AGAINST CHEMICAL AND BIOLOGICAL ATTACKS

Protective Equipment. Your main protection against a CB attack is your protective mask. It keeps you from inhaling chemical or biological agents. Additionally, protective clothing will provide protection from liquid agents. Protective clothing includes the mask with hood, the chemical protective suit (overgarment), boots, and gloves.

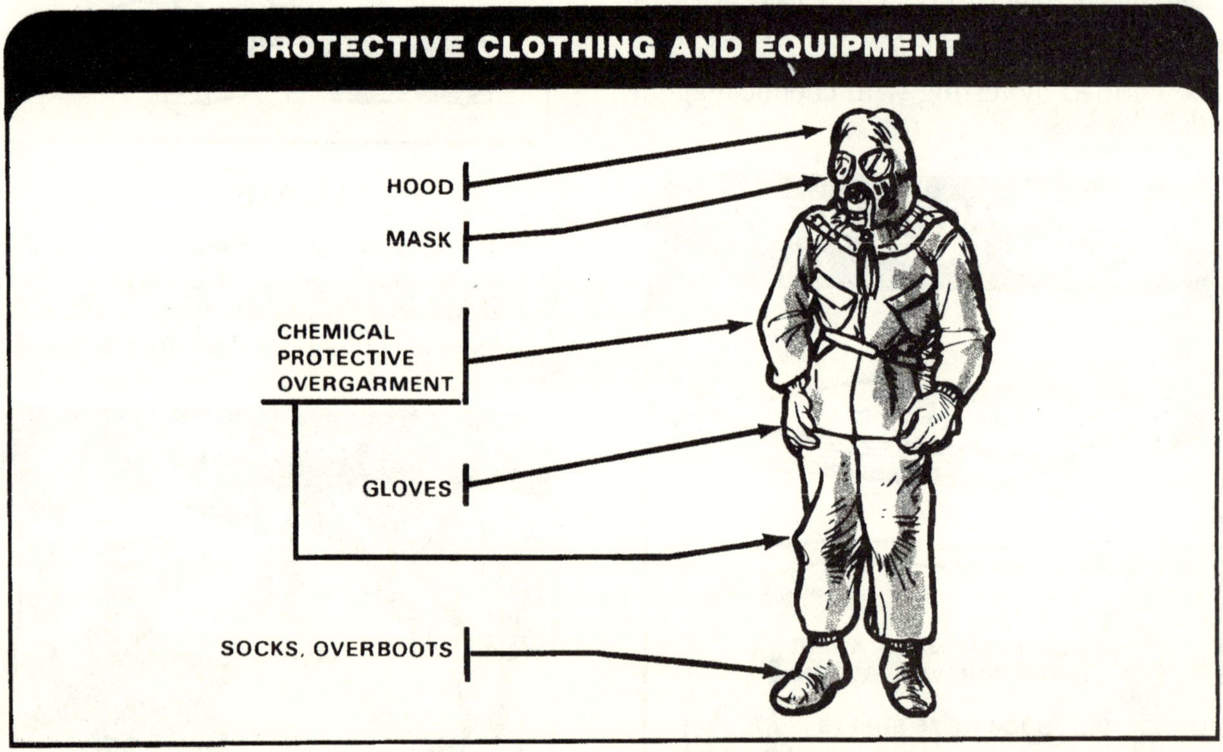

Protection from Insects. The duty uniform and gloves protect you against bites from insects such as mosquitoes and ticks that may carry disease-causing germs. Keep your clothes buttoned and your trouser legs tucked into your boots. Covering the skin reduces the chances of an agent entering the body through cuts and scratches. It also keeps disease-carrying insects from reaching the skin. Insect repellents and insecticides are effective against most disease-carrying insects. High standards of sanitation also protect against some insects.

Mission-Oriented Protective Posture. MOPP is a flexible system of protection against chemical agents. Your leader will specify the level of MOPP based on the chemical threat, workrate, and temperature prior to performing a mission. Later, he may direct a change in MOPP according to the changing situation.

The MOPP level determines what equipment you must wear and what you must carry. The standard MOPP levels are shown in the following chart.

MOPP LEVELS

	PROTECTIVE EQUIPMENT			
MOPP	OVERGARMENT	OVERBOOTS	MASK/HOOD	GLOVES
1	Worn opened or closed based on temperature.	Carried	Carried	Carried
2	Same as MOPP 1	Worn	Carried	Carried
3	Same as MOPP 1	Worn	Worn, hood opened or closed based on temperature	Carried
4	Worn, closed	Worn	Worn	Worn

The best local defense against biological warfare is strict preventive medical and sanitation measures and high standards of personal hygiene.

Chemical Attack. When an individual displays the symptoms of chemical-agent poisoning, first aid must be given immediately to save his life.

Nerve agents. The symptoms of nerve-agent poisoning are difficult breathing, drooling, nausea, vomiting, convulsions, and sometimes dim vision. The use of atropine autoinjectors and artificial respiration are first-aid measures for nerve-agent poisoning. If you have such symptoms, inject yourself with one injector in the thigh. If symptoms persist, use another injector. The interval between injections is 15 minutes. If you are unable to treat yourself, a buddy must do it for you. He will inject three injectors at once and administer artificial respiration, if necessary. No more than three atropine autoinjectors will be given. Seek medical aid quickly.

Blister agents. The symptoms of blister-agent poisoning are burning sensations in the skin, eyes, and nose. The symptoms may be immediate or delayed for several hours or days, depending on the type of agent used. If blister agents come in contact with the eyes or skin, decontaminate the areas at once. Decontaminate the eyes by flushing them repeatedly with plain water. Remove liquid blister agents from the skin by using the items of the M258A1 kit. If burns or blisters develop on the skin, cover them with sterile gauze or a clean cloth to prevent infection. Seek medical aid quickly.

Blood agents. The symptoms of blood-agent poisoning are nausea, dizziness, throbbing headache, skin/lips red or pink, convulsions, and coma. If those symptoms appear, hold two crushed ampules of amyl nitrite to the victim's nose. If in a contaminated area and the victim is wearing the protective mask, insert the crushed ampules inside the protective mask. If symptoms persist, repeat the treatment, using two crushed ampules about every 4 or 5 minutes until normal breathing returns or until eight ampules have been used. Breathing may become difficult or stop. Seek medical aid quickly.

Choking agents. The symptoms of choking-agent poisoning are coughing, choking, tightness of the chest, nausea, headache, and watering of the eyes. If you have these symptoms, stay quiet and comfortable, but seek medical aid quickly.

CHEMICAL-AGENT DECONTAMINATION OF SOLDIERS AND INDIVIDUAL EQUIPMENT

Use the M258A1 skin decontaminating kit to decontaminate your skim individual weapons, and equipment. Instructions for the use of the kit are printed on its container. This kit is especially made for skin decontamination however, you may use it to decontaminate some personal equipment such as your rifle, mask, and gloves.

The container for the M258A1 kit is a plastic waterproof case with a metal strap hook for attaching to clothing or equipment. It contains three Decon 1 wipes and three Decon 2 wipes, sealed in tear-away envelopes. Each Decon 1 wipe packet has a tab attached for night identification and to assist in removal from the case.

OPERATING INSTRUCTIONS FOR M258A1 KIT

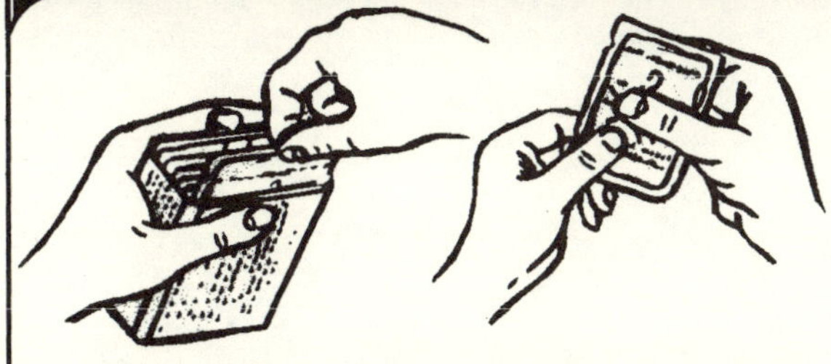

1. Pull out one Decon 2 packet. Crush inclosed glass ampules between them and fingers or smash glass ampules with palm of hand.

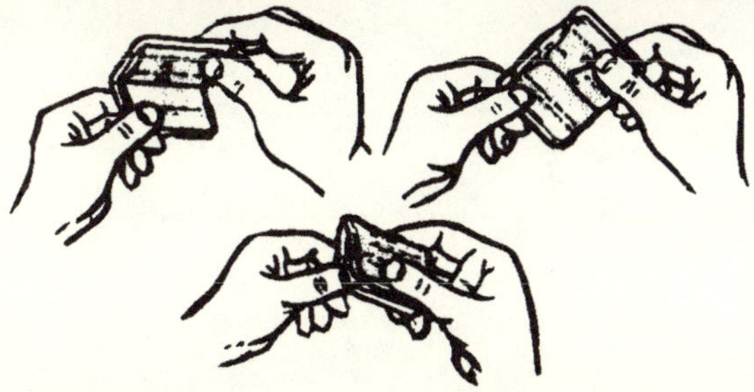

2. Fold packet on solid line marked crush and bend, then unfold.

3. Tear open quickly at notch and remove towelette.

4. Fully open towelette, let the encased crushed glass ampules fall away.

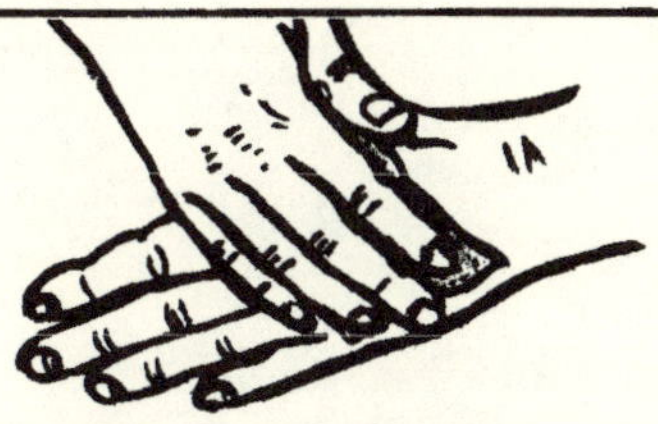

5. Wipe or swab exposed skin for 2 to 3 minutes. Start with hands then neck and ears. As necessary, bury the towelette and decon packets under several inches of soil.

CHEMICAL-AGENT DECONTAMINATION OF UNIT EQUIPMENT

Decontaminate key weapons with DS2 decontaminating solution, soapy water, solvents, or slurry. After decontamination, disassemble weapons and wash, rinse, and oil them to prevent corrosion. Decontaminate ammunition with DS2 solution, wipe with gasoline-soaked rags, and then dry it. If DS2 is not available, wash ammunition in cool, soapy water, then dry it thoroughly.

Decontaminate optical instruments by blotting them with rags, wiping with lens cleaning solvent, and then letting them dry.

Decontaminate communication equipment by airing, weathering, or hot air (if available).

BIOLOGICAL-AGENT DECONTAMINATION

Decontaminate your body by showering with soap and hot water. Use germicidal soap if available. Clean your nails thoroughly and scrub the hairy parts of your body. Wash contaminated clothing in hot, soapy water if it cannot be sent to a field laundry for decontamination. Cotton items may be boiled.

Wash vehicles with soapy water (prefer ably hot). If possible, steam-clean them using detergent.

Wash equipment in hot, soapy water and let it air dry.

CHAPTER 6

Combat Intelligence And Counterintelligence

GENERAL

Using the observation techniques discussed in chapter 4, you must **collect** and **report** information about the **enemy, terrain,** and **weather.** That information becomes combat intelligence after it is interpreted. Your leaders need combat intelligence to help them plan operations. Your life and the lives of your fellow soldiers could depend on reporting what you see, hear, and smell.

CONTENTS

GENERAL 6-1
SOURCES OF INFORMATION 6-2
WHAT TO REPORT 6-2
COUNTERINTELLIGENCE MEASURES 6-5

You must also act to keep the enemy from gaining information about US operations. That action, called **counterintelligence, involves:**

- **Denying the enemy information about US plans, intentions, and. activities.**
- **Detecting the enemy's efforts to get information.**
- **Deceiving the enemy as to US plans and intentions.**

SOURCES OF INFORMATION

Commanders get information from many agencies, but **you** are their best agency. You can collect information from the **following sources**

- **Prisoners of war (PW)** are an immediate source of information. Turn captured soldiers over to your leader quickly. Also, tell him anything you learn from them.
- **Captured documents** may contain valuable information about present or future enemy operations. Give such documents to your leader quickly.
- **Enemy activity** (the things the enemy is doing) often indicates what he is going to do. Report everything you see the enemy do. Some things that may not seem important to you may be important to your commander.
- **Local civilians** often have information about the enemy, terrain, and weather in an area. Report any information gained from civilians. However, you cannot be sure which side the civilians are trying to help, so be careful when acting on information obtained from them. Try to confirm that information by some other means.

WHAT TO REPORT

Report all information about the enemy to your leader quickly, accurately, and completely. Such reports should answer the questions WHO? WHAT? WHERE? after "WHEN?" It is best to use the "SALUTE" format (size, activity, location, unit, time, and equipment) when reporting. To help you remember details, make notes and draw sketches.

SALUTE

Size
Activity
Location
Unit
Time
Equipment

Size. Report the number of soldiers and vehicles you saw. For example, report " 10 enemy infantrymen" (not "a rifle squad") or "3 enemy tanks" (not "an enemy tank platoon").

Activity. Report what you saw the enemy doing. For example, "emplacing mines in the road."

Location. Report where you saw the enemy. If you have a map, try to give an eight-digit coordinate, such as "GL 874461." If you do not have a map, relate the location to some key terrain, such as "on the Harm Road, 300 meters south of the Ken River Bridge."

Unit. Report the enemy's unit. If the unit is not known, report any distinctive features, such as bumper markings on trucks, or type of headgear. Some armies have distinctive uniforms and headgear, or colored tabs on their uniforms, to identify types of units. A unit's action may also indicate its type. The kind of equipment observed may be peculiar to a certain type of unit. For example, a BRDM may indicate a reconnaissance unit.

Time. Report the time you saw the enemy activity, not the time you report it. Always report local or Zulu (Z) time.

Equipment. Report all of the equipment the enemy is wearing or using. If you do not recognize an item of equipment or a type of vehicle, sketch it and submit the sketch with the report. The following is an example of a SALUTE report.

FM: 1st Plt, C Co, 2d Bn, 1/73 Inf.
TO S2, 2d Bn, 1/73 Inf.

Combat OP sighted four enemy tanks moving west along secondary road at grid coordinates NB613397 at 241730Z. Tanks traveling at approximately 5 kilometers per hour. Hatches were open and visible enemy personnel were wearing protective masks.

PRISONERS OF WAR AND CAPTURED DOCUMENTS

PWs are a good source of information. They must be handled without breaking international law and without losing a chance to gain intelligence.

Treat PWs humanely. Do not harm them, either physically or mentally. The senior soldier present is responsible for their care. If PWs cannot be evacuated in a reasonable time, give them food, water, and first aid. Do not give them cigarettes, candy, or other comfort items. PWs who receive favors or are mistreated are poor interrogation subjects.

HANDLING PWs

In handling PWs, **follow the five S's:**

1. Search PWs as soon as they are captured. Take their weapons and papers, except identification papers and protective masks. Give them a written receipt for any personal property and documents taken. Tag documents and personal property to show which PW had them.

When searching a PW, have one man guard him while another searches him. (A searcher must not get between a PW and the guard.) To search a PW, have him spread-eagle against a tree or wall, or get into a pushup position with his knees on the ground. Search him, his equipment, his clothing.

2. Segregate PWs into groups by sex and into subgroups such as enlisted personnel, civilians, and political figures. This keeps the leaders from promoting escape efforts. Keep the groups segregated as you move them to the rear.

3. Silence PWs and do not let them talk to each other. This keeps them from planning escape and cautioning each other on security. Report anything a PW says or does.

4. Speed PWs to the rear. Turn them over to your leader. He will assemble them and move them to the rear for questioning by the S2.

5. Safeguard PWs when taking them to the rear. Do not let anyone abuse them. Watch out for escape attempts. Do not let PWs bunch up, spread out too far, or start diversions. Such conditions may create a chance for escape.

If a PW is wounded and cannot be evacuated through normal channels, turn him over to an aidman to be evacuated through medical channels.

Before evacuating a PW, attach a tag to him. You can make these tags yourself. The format for the tags is given in the following illustration. The battalion S2 should be able to supply these tags.

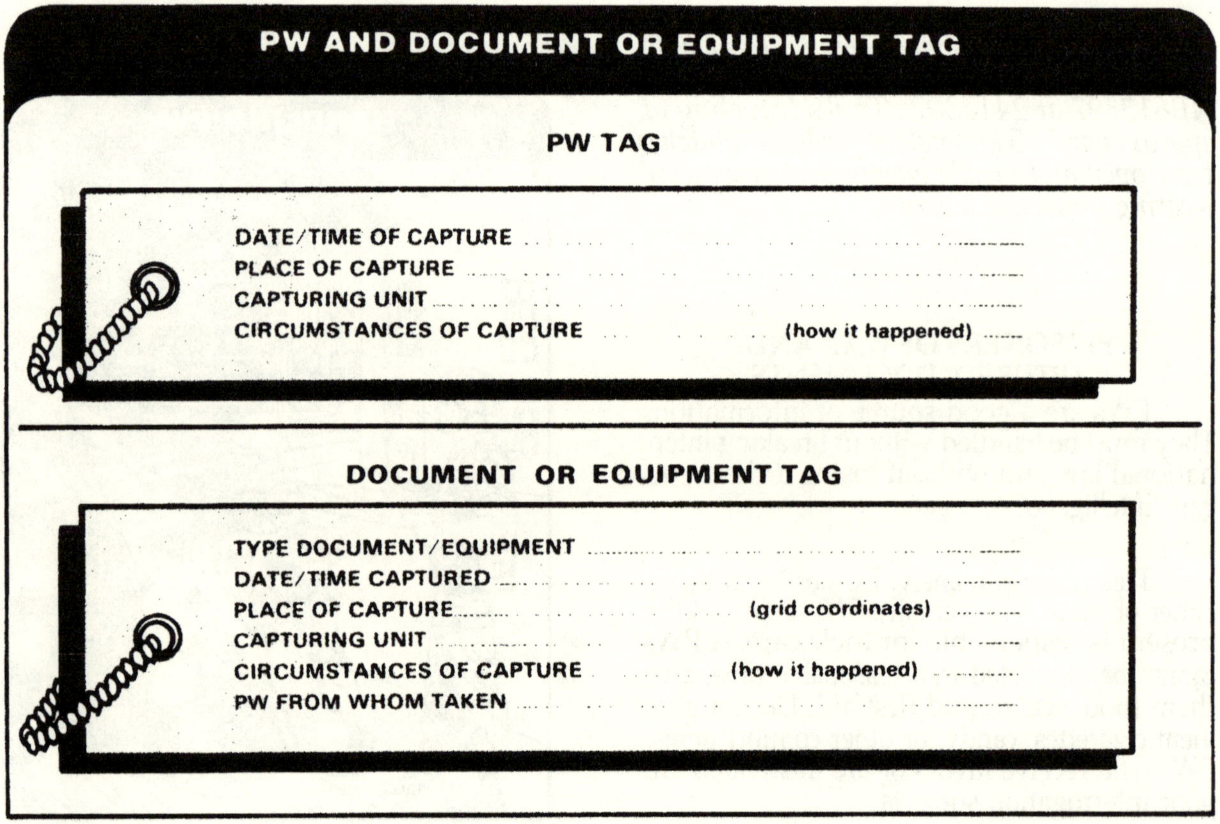

HANDLING CAPTURED DOCUMENTS AND EQUIPMENT

Enemy documents and equipment are good sources of information. Documents may be official (maps, orders, records, photos) or personal (letters or diaries).

If such items are not handled properly, the information in them may become lost or outdated. Give them to your leader quickly. Tag each item using the form shown above. If the item was found on a PW, put that PW's name on the tag.

COUNTERINTELLIGENCE MEASURES

The enemy must not get information about US operations. This means that you and your **fellow soldiers must:**

- **Practice camouflage principles and techniques.**
- **Practice noise and light discipline.**
- **Practice field sanitation.**
- **Use proper radiotelephone procedure.**
- **Use the challenge and password properly.**
- **Not take personal letters or pictures into combat areas.**
- **Not keep diaries in combat areas.**
- **Be careful when discussing military affairs (the enemy may be listening).**
- **Use only authorized codes.**
- **Abide by the Code of Conduct (if captured).**
- **Report any soldier or civilian who is believed to be serving or sympathetic with the enemy.**
- **Report anyone who tries to get information about US operations.**
- **Destroy all maps or important documents if capture is imminent.**
- **Not discuss military operations in public areas.**
- **Discuss military operations only with those persons having a need to know the information.**
- **Remind fellow soldiers of their counterintelligence responsibilities.**

CHAPTER 7

Communications

GENERAL

Communications are exchanges of information by two or more parties. The information must be transmitted and received/understood.

You must know how to communicate with your leaders and fellow soldiers. **You must be able to tell:**

- **What you see.**
- **What you are doing.**
- **What you have done.**
- **What you are going to do.**
- **What you need.**

CONTENTS

GENERAL 7-1
MEANS OF COMMUNICATION 7-2
RADIOTELEPHONE PROCEDURE 7-4
COMMUNICATIONS SECURITY 7-10
RADIO EQUIPMENT 7-11
WIRE AND TELEPHONE EQUIPMENT 7-15

MEANS OF COMMUNICATION

There are several means of communications. Each has its own capabilities, advantages, and disadvantages. Those you can use are described in this section.

RADIO COMMUNICATIONS

Radios are a frequently used means of communications. Radios are particularly suited for use when you are on the move and need a means of maintaining command and control. Small handheld or backpacked radios that communicate for only short distances are found at squad and platoon level. As the need grows to talk over greater distances and to more units, the size and complexity of radios are increased.

To put these radios to good use, you must first look at some of the things that affect radio communications. To communicate with each other, radios must have a common frequency. They must also be able to transmit and receive the same type signal. Most infantry radios are FM (frequency modulated) and will not communicate with AM (amplitude modulated) radios. Squelch settings on the radios must also be used correctly.

Factors that affect the range of radio equipment are weather, terrain, antenna, power, and the location of the radio. Trying to communicate near man-made objects such as bridges and buildings may also affect radio transmis-

MEANS OF COMMUNICATION

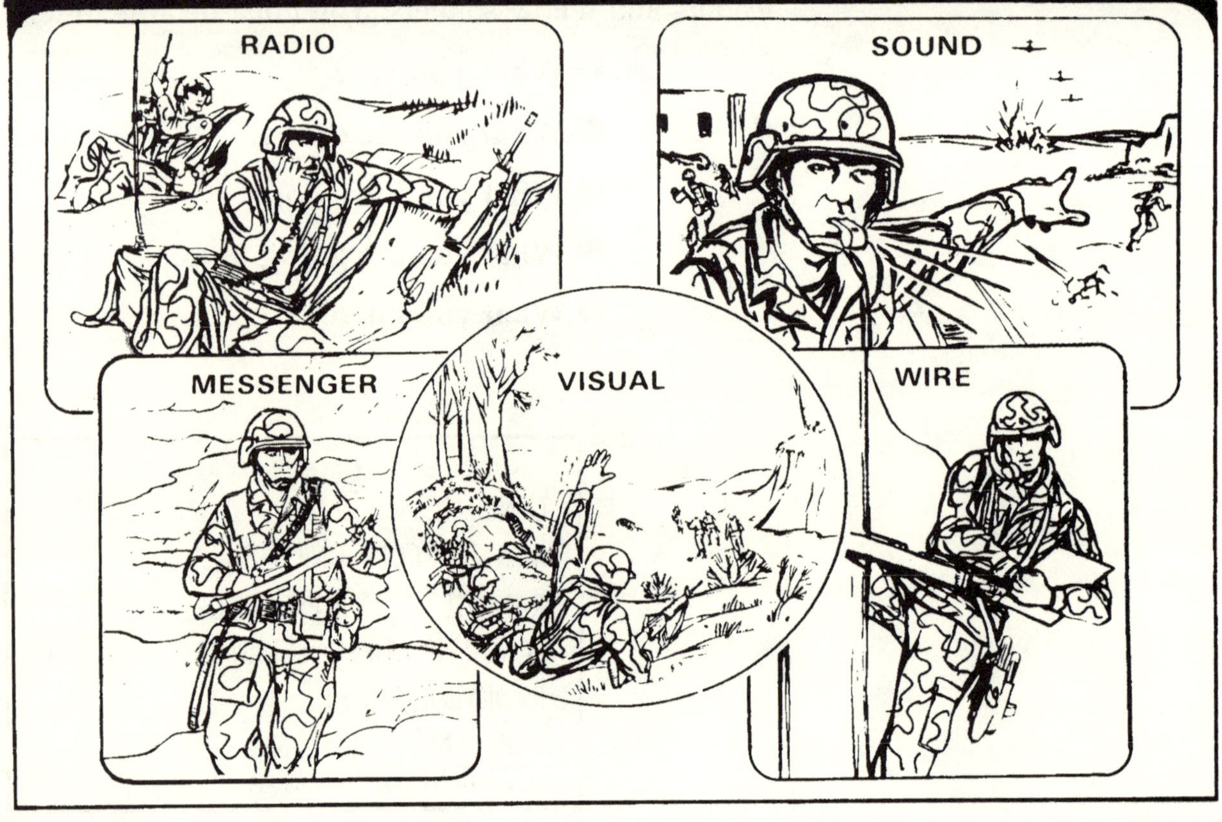

sions. Interference in the form of static often occurs when you use radios near powerlines or electrical generators. Interference may also come from other radio stations, bad weather, or enemy jamming.

Many of the things that may cause poor radio communications can be corrected by using common sense. Such things as making sure that you are not trying to communicate from under a steel bridge or near generators and powerlines, using the best available antenna for your needs, and selecting the best site for your radio help insure more reliable communications. You can also reduce the effects of enemy jamming by employing antijamming techniques.

Radio is one of the least secure means of communicating. Each time you talk over a radio, the sound of your voice travels in all directions. The enemy can listen to your radio transmissions while you are communicating with other friendly radio stations. You must always assume that the enemy is listening to get information about you and your unit, or to locate your position to destroy you with artillery fire. Everyone who uses radios must know the defensive techniques available to prevent the enemy from getting information.

VISUAL COMMUNICATIONS

The enemy's ability to interfere with your radio signals is causing more emphasis to be placed on visual communications for command and control. Visual signals include arm-and-hand signals, pyrotechnics, smoke, flashing lights, panel markers, and aircraft maneuvers.

The effectiveness of any visual signal depends on a set of prearranged meanings. You assign prearranged meanings to visual signals to the soldier sending the signal and the soldier seeing the signal so both have the same understanding of what that particular signal means. Your commander will set prearranged meanings for pyrotechnics, smoke, and flashing lights. Generally, a listing of prearranged messages using these signals is contained in your unit SOP or communications-electronics operation instructions (CEOI).

Panel markers are a series of cloth panels that you spread on the ground to communicate with aircraft. They are useful when you do not have radio contact with friendly aircraft, when ground units and/or aircraft are on radio listening silence, when your radio equipment has been damaged or destroyed, or when enemy jamming makes radio communications difficult or impossible. When standard cloth panels are not available, you can use field expedients such as clothing, branches, rocks, or snow.

Panel codes, as well as arm-and-hand signals, have standard prearranged meanings. The prearranged meaning of arm-and-hand signals may be found in FM 21-60. Information is usually taken from those publications and placed in unit CEOIs and SOPs.

Visual signals have some shortcomings that limit their use. For example, visual signals can be easily misunderstood. Some visual signals are restricted during poor visibility such as at night or in dense terrain. Of course, at other times, they can be intercepted by the enemy who may, in turn, use similar visual signals to create confusion.

SOUND COMMUNICATIONS

Sound signals, like visual signals, depend upon a set of prearranged meanings. Sound signals include the use of the voice, whistles, horns, weapons, and other noise-making devices to transmit simple messages over short distances. Also, like visual signals, sound signals are vulnerable to enemy interception and use. Battle noise can obviously reduce the effective use of sound signals. They have their greatest application as command post warning alarms. The prearranged meanings for sound signals are usually established by local commanders, and a listing of such meanings is commonly found in unit SOPS and the CEOIs. Sound signals, like visual signals, can be easily misunderstood.

WIRE COMMUNICATIONS

Wire is another type of communications used in infantry units. Although installing a wire network takes more time than installing a radio, wire lines are usually more secure than radio. When you talk over wire lines, your voice travels through the wire lines from one telephone to another and is generally not sent through the air. Wire lines will give better communications in most cases because they are less subject to interference from weather, terrain, and man-made obstacles. Wire lines also protect you from enemy electronic warfare actions such as jamming.

Wire lines are subject to breakage by enemy artillery and air strikes and by friendly forces who accidentally cut the lines when driving over them with tracked and wheeled vehicles. It is important, therefore, to install wire lines properly to reduce the possibility of breakage. When laying wire lines, first consider the tactical situation. In a fast-moving situation, the use of wire may be impractical. In a static situation, you have more time to install wire lines.

Consideration must be given to the enemy's ability to jam radios and to locate positions through direction finding when you communicate by radio. If the enemy has displayed such capabilities, wire should be considered as an alternative to radio. The terrain will also influence use of wire communications. Wire laying may be difficult in dense vegetation, in swampy areas, or in mountainous terrain. Rain, snow, and temperature extremes may also influence wire laying. Men and equipment to lay wire lines should be available.

MESSENGER COMMUNICATIONS

Unlike other infantry communications, messengers are a means of transmitting large maps, documents, and bulk material, as well as oral or written messages. Message centers serve as a central point for receiving and distributing message-type information. They are located at battalion or higher level headquarters. Messenger service may be limited, however, because messengers are subject to enemy action, require more time than radio or wire communications, and do not afford real time writer-to-reader exchanges.

RADIOTELEPHONE PROCEDURE

Radiotelephone procedure is a set procedure for using a radio or telephone. It speeds the exchange of messages and helps avoid errors. The rules listed below will help you use transmission times efficiently and avoid violations of communications security.

1. Transmit clear, complete, and concise messages. When possible, write them out beforehand.

2. Speak clearly, slowly, and in natural phrases. Enunciate each word. If a receiving operator must write the message, allow time for him to do so.

3. Listen before transmitting, to avoid interfering with other transmissions.

4. ALWAYS ASSUME THE ENEMY IS LISTENING.

PHONETIC ALPHABET

To help identify spoken letters, a set of easily understood words has been selected. Those words help to avoid confusion. BRAVO, for example, is the phonetic word of the letter B, and DELTA is the phonetic word for the letter D. BRAVO and DELTA are less likely to be confused in a radio message than B and D. **Use the phonetic alphabet to:**

- **Transmit isolated letters.**
- **Transmit each letter of an abbreviation.**
- **Spell out unusual or difficult words.**

PRONUNCIATION OF WORDS

LETTER	WORD	SPOKEN AS	LETTER	WORD	SPOKEN AS
A	ALPHA	**AL** FAH	N	NOVEMBER	NO **VEM** BER
B	BRAVO	**BRAH** VOH	O	OSCAR	**OSS** CAH
C	CHARLIE	**CHAR** LEE/ **SHAR** LEE	P	PAPA	PAH **PAH**
			Q	QUEBEC	KEH **BECK**
D	DELTA	**DELL** TAH	R	ROMEO	**ROW** ME OH
E	ECHO	**ECK** OH	S	SIERRA	SEE **AIR** RAH
F	FOXTROT	**FOKS** TROT	T	TANGO	**TANG** GO
G	GOLF	GOLF	U	UNIFORM	**YOU** NEE FORM/ **OO NEE FORM**
H	HOTEL	HOH **TELL**			
I	INDIA	**IN** DEE AH	V	VICTOR	**VIK** TAH
J	JULIETT	JEW LEE **ETT**	W	WHISKEY	**WISS** KEY
K	KILO	**KEY** LOH	X	X-RAY	**ECKS** RAY
L	LIMA	**LEE** MAH	Y	YANKEE	**YANG** KEY
M	MIKE	MIKE	Z	ZULU	**ZOO** LOO

(NOTE: Syllables in bold print carry the accent).
When you must spell out a difficult word in the text of a message, precede it by the proword "I SPELL." If you can pronounce the word, do so before and after spelling it.

Example: The word MANEUVER must be transmitted and can be pronounced. "MANEUVER – I SPELL - Mike-Alpha-November-Echo -Uniform-Victor- Echo-Romeo – MANEUVER."

If you cannot pronounce the word, do not attempt to pronounce it. Instead, precede the word with the proword "I SPELL."

Example The word EVACUATE must be transmitted and cannot be pronounced. "I SPELL - Echo-Victor-Alpha-Charlie-Uniform-Alpha-Tango-Echo."

PRONUNCIATION OF NUMBERS

NUMERAL	SPOKEN AS
0	ZE-RO
1	WUN
2	TOO
3	TREE
4	FOW-ER
5	FIFE
6	SIX
7	SEV-EN
8	AIT
9	NIN-ER

Transmit multiple digit numbers digit by digit. Two exceptions to this are when transmitting exact multiples of thousands and when identifying a specific code group in a coded message. When calling for or adjusting field artillery or mortar fire, it is necessary to transmit, when applicable, exact multiples of hundreds and thousands using the appropriate noun.

MULTIPLE DIGIT NUMBERS	
NUMBER	SPOKEN AS
44	FOW-ER FOW-ER
90	NIN-ER ZE-RO
136	WUN TREE SIX
500	FIFE ZE-RO ZE-RO
1,200	WUN TOO ZE-RO ZE-RO
1,478	WUN FOW-ER SEV-EN AIT
7,000	SEV-EN TOU-SAND
16,000	WUN SIX TOU-SAND
812,681	AIT WUN TOO SIX AIT WUN

PROWORDS

Certain procedural words (prowords) which have distinct meaning should be used to shorten transmissions and avoid confusion.

PROWORDS AND THEIR EXPLANATIONS	
PROWORDS	EXPLANATION
ALL AFTER	The part of the message to which I refer is all of that which follows.
ALL BEFORE	The part of the message to which I refer is all of that which precedes.
AUTHENTICATE	The station called is to reply to the challenge which follows.
AUTHENTICATION IS	The transmission authentication of this message is ______________.
BREAK	I hereby indicate the separation of the text from other parts of the message.
CORRECT	You are correct, or what you have transmitted is correct.

PROWORD	EXPLANATION
CORRECTION	An error has been made in this transmission. Transmission will continue with the last word correctly transmitted. An error has been made in this transmission (or message indicated). The correct version is ______________. That which follows is a corrected version in answer to your request for verification.
FLASH	Flash precedence is reserved for alerts, warnings, or other emergency actions having immediate bearing on national, command, or area security (e.g., presidential use; announcement of an alert; opening of hostilitie; land, air, or sea catastrophles; intelligence reports on matters leading to enemy attack; potential or actual nuclear accident or incident; implementation of services unilateral emergency actions procedures).
FROM	The originator of this massage is indicated by the address designator immediately following.
GROUPS	This message contains the number of groups indicated by the numeral following.
I AUTHENTICATE	The group that follows is the reply to your challenge to authenticate.
IMMEDIATE	Immediate precedence is reserved for vital corn. munications that (1) have an immediate operational effect on tactical operations, (2) directly concern safety or rescue operations, (3) affect the intelligence community operational role (e.g., initial vital reports of damage due to enemy action; land, sea, or air reports that must be completed from vehicles in motion such as operational mission aircraft; intelligence reports on vital actions in progress; natural disaster or widespread damage; emergency weather reports having an immediate bearing on mission in progress; emergency use for circuit restoration; use by tactical command posts for passing immediate operational traffic).
I READ BACK	The following is my response to your instructions to read back.

PROWORD	EXPLANATION
I SAY AGAIN	I am repeating transmission or part indicated.
I SPELL	I shall spell the next word phonetically.
MESSAGE	A message which requires recording is about to follow. Transmitted immediately after the call. (This proword is not used on nets primarily employed for conveying messages. H is intended for use when messages are passed on tactical or reporting nets.)
MORE TO FOLLOW	Transmitting station has additional traffic for the receiving station.
OUT	This is the end of my transmission to you and no answer is required.
OVER	This is the end of my transmission to you and a response is necessary. Go ahead: transmit.
PRIORITY	Priority precedence is reserved for calls that require prompt completion for national defense and security, the successful conduct of war, or to safeguard life or property, and do not require higher precedence (e.g., reports of priority land, sea, or air movement; administrative, intelligence, operational or logistic activity calls requiring priority action; calls that would have serious impact on military, administrative, intelligence, operational, or logistic activities if handled as a ROUTINE call). Normally, PRIORITY will be the highest precedence that may be assigned to administrative matters for which speed of handling is of paramount importance.
RADIO CHECK	What is my signal strength and readability. In other words, how do you read (hear) me?
READ BACK	Repeat this entire transmission back to me exactly as received.
RELAY	Transmit this message to all addressees immediately following this proword.
ROGER	I have received your last transmission satisfactorily, and loud and clear.

PROWORD	EXPLANATION
ROUTINE	Routine precedence is reserved for all official communications that do not require flash, immediate, or priority precedence.
SAY AGAIN	Repeat your last transmission or the part indicated
SILENCE	Cease transmissions on this net immediately.
(Repeated three or more times.)	Silence will be maintained until lifted. (When an authentication system is in force, the transmission imposing silence is to be authenticated.)
SILENCE LIFTED	Silence is lifted. (When an authentication system is in force, the transmission lifting silence is to be authenticated.)
SPEAK SLOWER	You are transmitting too fast. Slow down.
THIS IS	This transmission is from the station whose designation immediately follows.
TIME	That which immediately follows is the time or date-time group of the message.
TO	The addressees immediately following are addressed for action.
UNKNOWN STATION	The identity of the station with whom I am attempting to communicate is unknown.
WAIT	I must pause for a few seconds.
WAIT-OUT	I must pause longer than a few seconds.
WILCOX	I have received your signal, understand It, and will comply. To be used only by the addressee. As the meaning of ROGER is included in that of WILCO, the two prowords are never used together.

COMMUNICATIONS SECURITY

Communications security keeps unauthorized persons from gaining information of value from radio and telephone transmissions. It includes:

- **Using authentication to make sure that the other communicating station is a friendly one.**
- **Using only approved codes.**
- **Designating periods when all radios are turned off.**
- **Restricting the use of radio transmitters and monitoring radio receivers.**
- **Operating radios on low power.**

- Enforcing net discipline and radio-telephone procedure (all stations must use authorized prosigns and prowords, and must transmit official traffic only).
- Using radio sites with hills or other shields between them and the enemy.
- Using directional antennas when feasible.

RADIO EQUIPMENT

soldier should be familiar with the **AN/PRC-77 radio** and the two types of squad radios. One type of squad radio is the **AN/PRC-68 Small Unit Transceiver (SUT).** The other comes in two parts the **AN/PRT-4 (transmitter)** and the **AN/PRR-9 (receiver).**

To operate the **AN/PRC-77 radio:**

- Install the battery.
- Replace the battery compartment and close both latches at the same time.
- Select the antenna (plus antenna base) and tighten it down.
- Connect the handset.
- Select the frequency band.
- Set the frequency using the tuning control knobs.
- Turn the function switch to ON.
- Turn the volume control knob about half a turn.
- Depress the push-to-talk switch on the handset to talk and release it to listen.
- Adjust the volume control to the desired level.

RADIO AN/PRC-77

The **AN/PRC-77 RADIO** set has a planning range of 5 to 8 km and weighs 12 kg **(24.7 lb)** with batteries. The battery life is 60 hours with BA-4386.

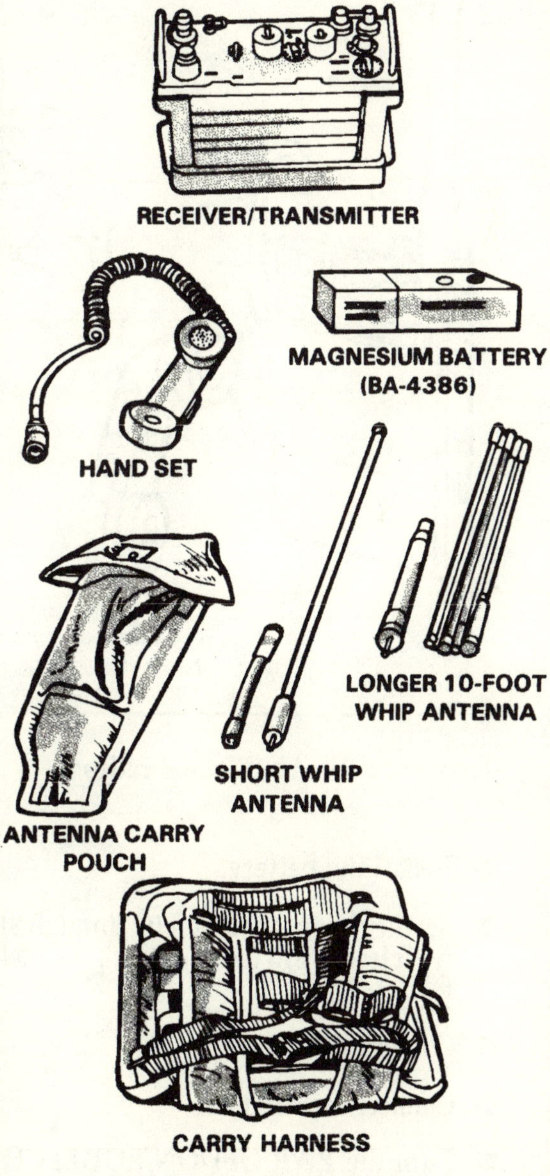

RADIO AN/PRC-68

The **AN/PRC-68 RADIO (Small Unit Transceiver [SUT])** has a planning range of 1 to 3 km and weighs .99 kg (35 oz). It has 1000 channels and can be preset on 10 of them. Its battery's life is 24 hours.

The battalion communications platoon presets the channels.

To operate the **AN/PRC-68 radio**

- **Install the battery.**
- **Set the channel position (O through 9) (your leader will tell you which channel to use).**
- **Connect the handset.**
- **Connect the antenna.**
- **Turn the PWR OFF/ON/SQUELCH switch to ON.**
- **Turn the PWR OFF/ON/SQUELCH switch to SQUELCH (this switch is spring-loaded and will return to ON when disengaged).**
- **Turn the volume knob to adjust loudness of received signal.**
- **Depress the push-to-talk switch on the handset or the back of the radio to transmit and release it to listen.**

NOTE: Handset is not a part of AN/ PRC-68 (H-189 and H-250 handset).

RADIO AN/PRR-9

The receiver **(AN/PRR-9)** will receive Channel 1 and Channel 2, but not both at the same time. It is also battery-powered. Two types of batteries can be used in the receiver. The dry-cell battery **(BA-505/U)** has a life of about 14 hours. The magnesium battery **(BA-4504/U)** has a life of about 28 hours.

To operate the **AN/PRR-9 radio**

- **Insert the tubular BA-505/U battery through the battery clip and into the mating connector of the receiver.**
- **Clip the receiver to your helmet.**
- **Loosen the antenna retaining screw and rotate the antenna upright (retighten retaining screw).**
- **Set the receiver control.**

For receiving with squelch, turn the receiver control knob clockwise from its OFF position. Set it to a comfortable listening level when voice or tone is heard. If the control knob is turned clockwise to its last position, the squelch is turned off (background noise will start). To reactivate the squelch, turn the control knob to OFF, then back about halfway toward ON.

For receiving without squelch, turn the receiver control knob fully clockwise from its OFF position. Turn it counterclockwise to a comfortable listening level. Do not use squelch when signals are weak or in terrain unfavorable for good reception.

Wear the receiver either on your combat suspenders or clipped to your pocket, belt, or helmet. Use a lanyard to tie the receiver down.

AN/PRT-4 RADIO

The transmitter (AN/PRT-4) is battery-powered and has two channels. Channel 1 has a range of 1,600 meters. Channel 2 has a range of 500 meters. The AN/PRT-4 can transmit a tone as well as voice. Battery life is about 35 hours for the BA-399.

To operate the **AN/PRT-4 radio:**

- **Release both battery case clamps and remove the battery case.**
- **Insert a BA-399/U battery into the mating connector at the bottom of transmitter.**
- **Replace the battery case and secure the clamps.**
- **Raise the collapsible antenna to its full height.**
- **Set the upper selector switch in the CH-1 position for channel 1 or the CH-2 position for channel 2.**
- **Set the TONE-VOICE switch.**

For a tone signal, turn the tone-voice switch to the TONE position and hold it in that position for as long as the tone signal is needed. Release the switch at the end of that time.

For voice communications, turn the tone-voice switch to the VOICE position and hold it in that position while transmitting. Speak into the microphone located above the channel selector switch. Release the tone-voice switch at the end of the transmission.

To permit transmissions in only one mode, position the override spring on either VOICE or TONE, depending on which is needed.

Wear the transmitter clipped to your pocket, belt, or suspenders. To prevent loss of the transmitter, use a lanyard to tie it down.

WIRE AND TELEPHONE EQUIPMENT

When in the defense, units normally communicate by wire and messenger instead of by radio. Your leaders will often have you lay the wire and install and operate the field phones.

WIRE-LAYING TECHNIQUES

A surface line is field wire laid on the ground. Lay surface lines loosely with plenty of slack. Slack makes installation and maintenance easier. Surface lines take less time and fewer soldiers to install. When feasible, dig small trenches for the wire to protect it from shell fragments of artillery or mortar rounds. Conceal wire routes crossing open areas from enemy observation. Tag all wire lines at switchboards and at road, trail, and rail crossings to identify the lines and make repair easier if a line is cut.

An **overhead line** is field wire laid above the ground. Lay overhead lines near command posts, in assembly areas, and along roads where heavy vehicular traffic may drive off the road. Also, lay them at road crossings where trenches cannot be dug, if culverts or bridges are not available. Those lines are the least likely to be damaged by vehicles or weather.

DR-8

The DR-8 holds 400 meters of field wire (WD-1). The wire is reusable and should be taken up if the situation permits.

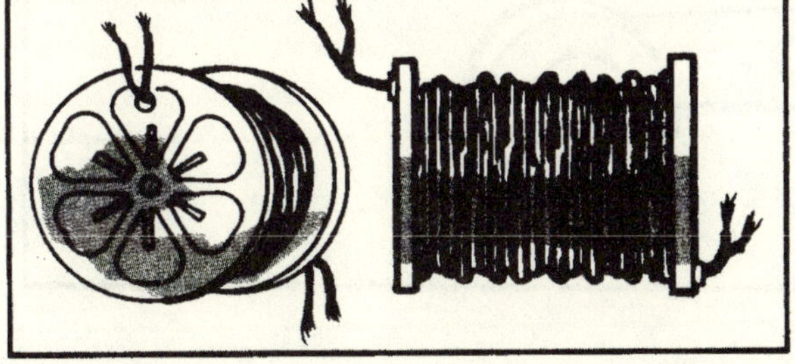

The telephone set TA-1 is a sound-powered phone that has both a visual and an audible signal. It has a range of 6.4 km using WD-1 wire.

TA-1 TELEPHONE

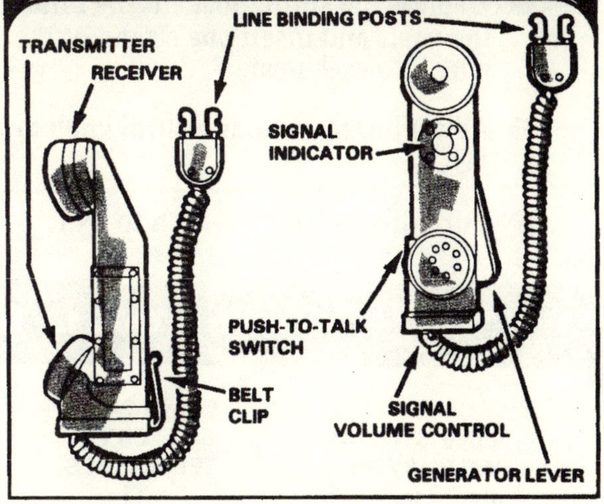

To install the TA-1 telephone:

- **Strip away half an inch of insulation from each strand of the WD-1 wire line.**
- **Depress the spring-loaded line binding posts and insert one strand of the wire into each post.**
- **Adjust signal volume control knob to LOUD.**
- **Depress the generator lever several times to call the other operator and listen for buzzer sound.**
- **Turn the buzzer volume control knob until the wanted volume is obtained.**
- **Look at the visual indicator to see if it shows four white luminous markings.**
- **Depress the push-to-talk switch to reset the visual indicator.**

The telephone set TA-312 is a battery-powered phone. It has a range of 38 km using WD-1 wire.

To install the TA-312 telephone:

- **Strip away one-half inch of insulation from each strand of the WD-1 wire line.**
- **Depress the spring-loaded line binding posts and insert one strand of the wire into each post.**
- **Adjust buzzer volume control knob to LOUD.**
- **Turn the INT-EXT switch to INT.**
- **Turn the circuit selector switch to LB.**
- **Insert the two BA-30 batteries into the battery compartment (one up and one down).**
- **Seat the handset firmly in the retaining cradle.**
- **Turn the handcrank rapidly a few turns. Remove the handset from the retaining cradle and wait for the other operator to answer.**
- **Depress the push-to-talk switch to talk. Release the push-to-talk switch to listen.**

TA-312 TELEPHONE

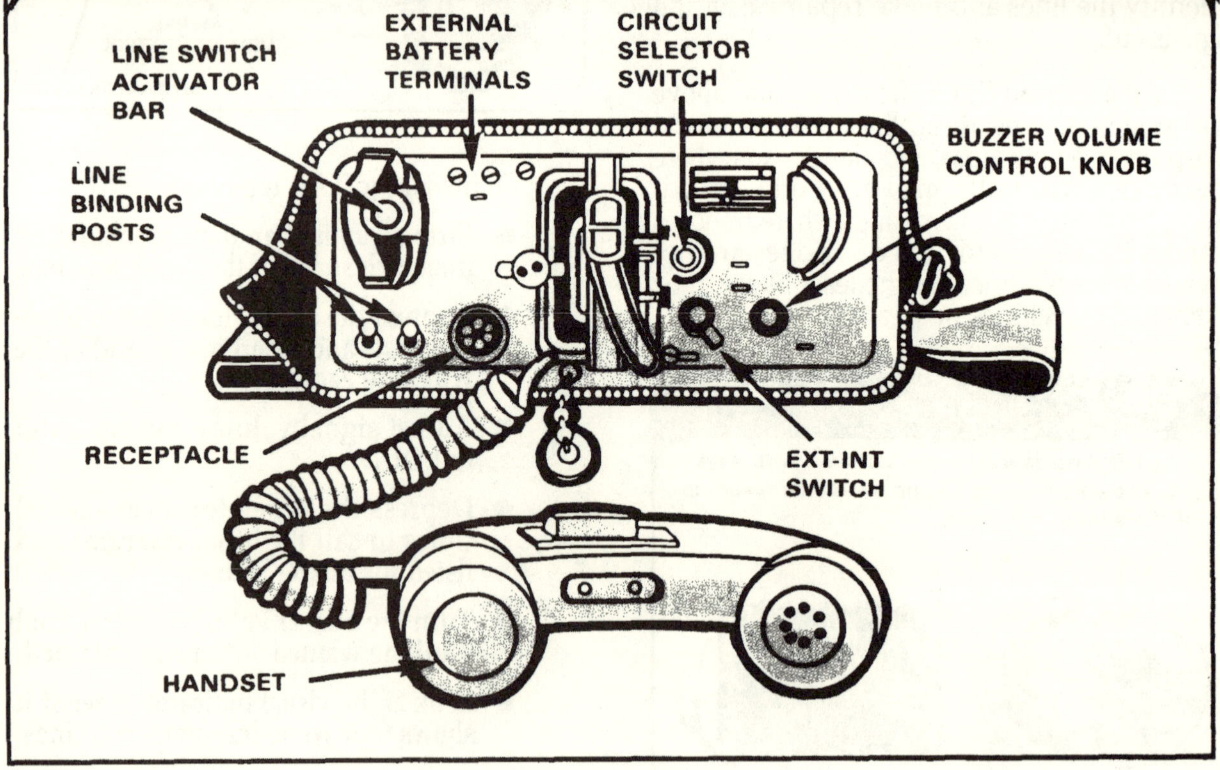

REEL EQUIPMENT CE-11

The reel equipment CE -11 is a lightweight, portable unit used for laying and picking up short wire lines. **It has the following components:**

- **Reeling machine, cable, band, RL-39, with an axle and crank, carrying handles, and straps ST-34 and ST-35.**
- **Telephone set TA-1/PT.**
- **The RL-39 component mounts the reel cable DR-8 that will hold 400 meters of field wire WD-1/TT. The DR-8 and the wire are separate items and ARE NOT part of the CE-11 or the RL-39.**

The major parts of the CE -11 may also be authorized by TOE as separate items and not as a complete unit CE -11.

REEL EQUIPMENT, CE-11

The **Reel Equipment CE-11** is a light, portable unit for laying and picking up short wire lines. **It has the following components:**

- **Telephone Set TA-1**

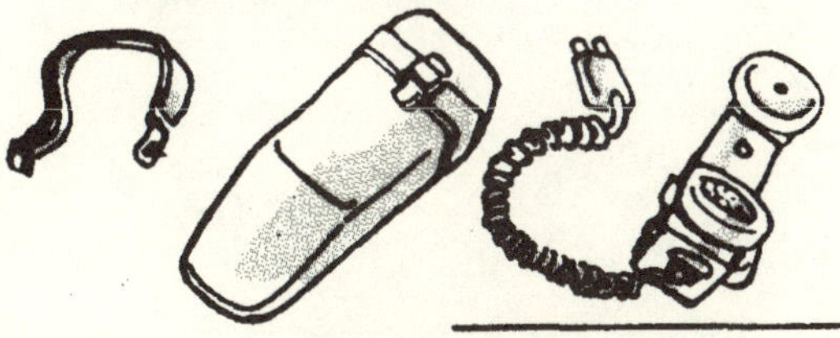

- **Reeling Machine Cable Hand RL-39** with an axle and crank, carrying handles and straps. It mounts the DR-8 reel cable **(below)** which is a separate item.

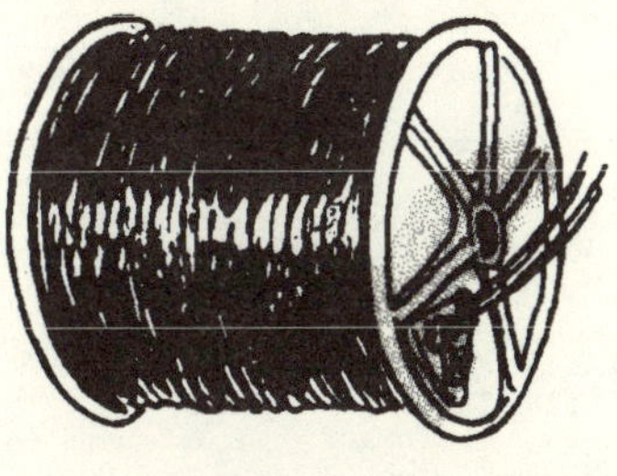

The DR-8 holds 400 meters of field wire **(WD-1)**. The wire is reusable and should be taken up if the situation permits.

CHAPTER 8

First Aid And Personal Hygiene

GENERAL

First aid is the care and treatment you give a casualty before medical personnel arrive. Personal hygiene is the steps you take to protect your own health and that of others. Your personal-hygiene and first-aid skills could save your life or the life of a buddy.

By knowing what to do, and by getting medical help quickly, you may be able to save lives, prevent permanent disabilities, and prevent long periods of hospitalization.

CONTENTS

GENERAL 8-1
LIFESAVING MEASURES 8-2
DOs AND DON'Ts OF FIRST AID 8-16
PERSONAL HYGIENE 8-16

The field first-aid packet issued to you should be carried at all times for personal use. It contains one or two field first-aid dressings. Use the first-aid dressings on wounds. When giving first aid to a casualty, you should use the casualty's first-aid items. You may need your own items later if you become injured. For more information on first aid, **see FM 21-11.**

FIRST AID CASE AND COMPONENTS

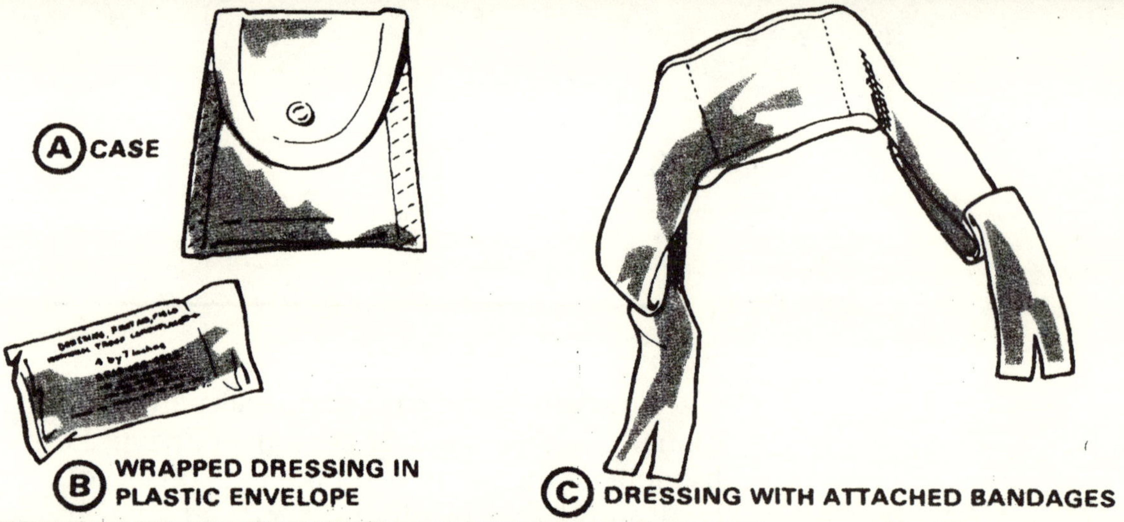

LIFESAVING MEASURES

When you or your buddy is wounded, first aid must be given at once. The first step is to apply (as needed) the four life-saving measures. **These measures are:**

- **Clear the airway; cheek and restore breathing and heartbeat.**
- **Stop the bleeding.**
- **Prevent shock.**
- **Dress and bandage the wound.**

CLEAR THE AIRWAY; CHECK AND RESTORE BREATHING AND HEARTBEAT

Clear the Airway. The lack of oxygen intake through breathing and lack of heartbeat leads to death in a very few minutes.

When treating a casualty, first find out if he is breathing. **If he is not breathing**

- **Place him on his back and kneel beside his head.**
- **Clear his airway by removing any obstruction in his mouth.**
- **Place your hand (the hand nearest his feet) under his neck and put your other hand on his forehead. Extend his neck by lifting with the hand under the neck and pushing down on the forehead. This also lifts the tongue away from the back of the throat, opening the airway.**

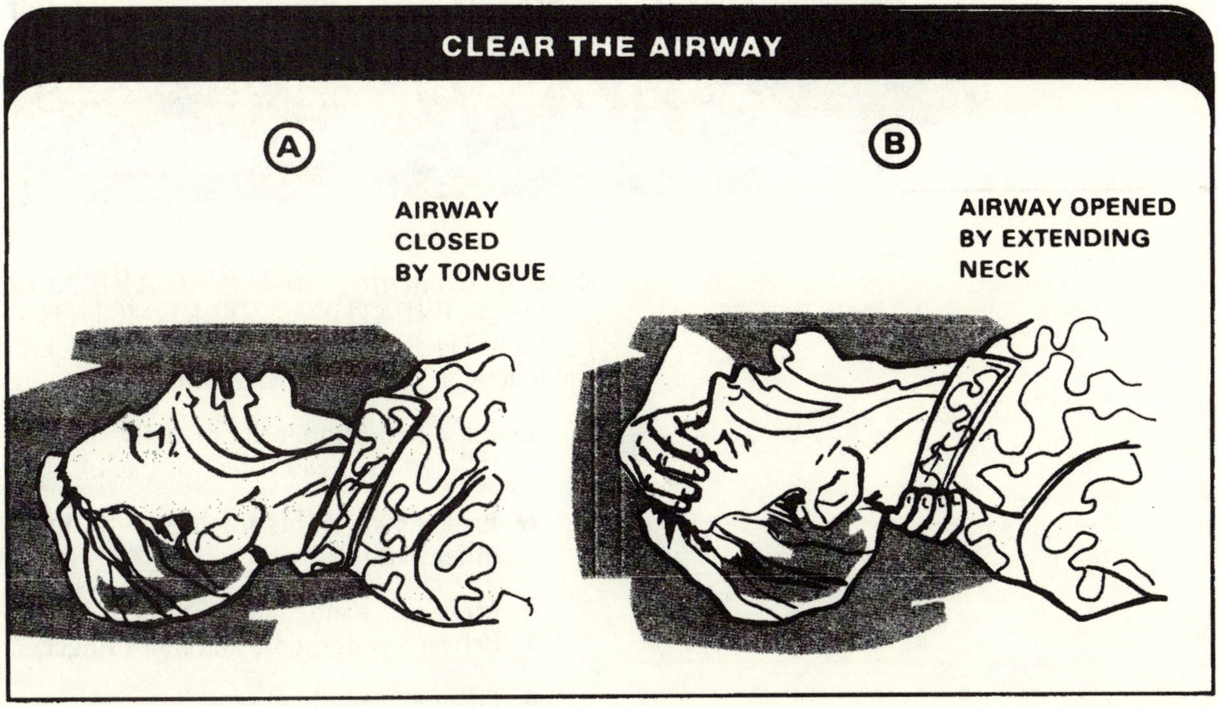

Check for Breathing. After opening the airway, LOOK, LISTEN, and FEEL to find out if the casualty is breathing. The following procedures **should be used:**

- **Put your ear near the casualty's mouth and nose: hold this position for about 5 seconds.**
- **LOOK to see if the casualty's chest is rising and falling.**
- **LISTEN and FEEL for breathing.**

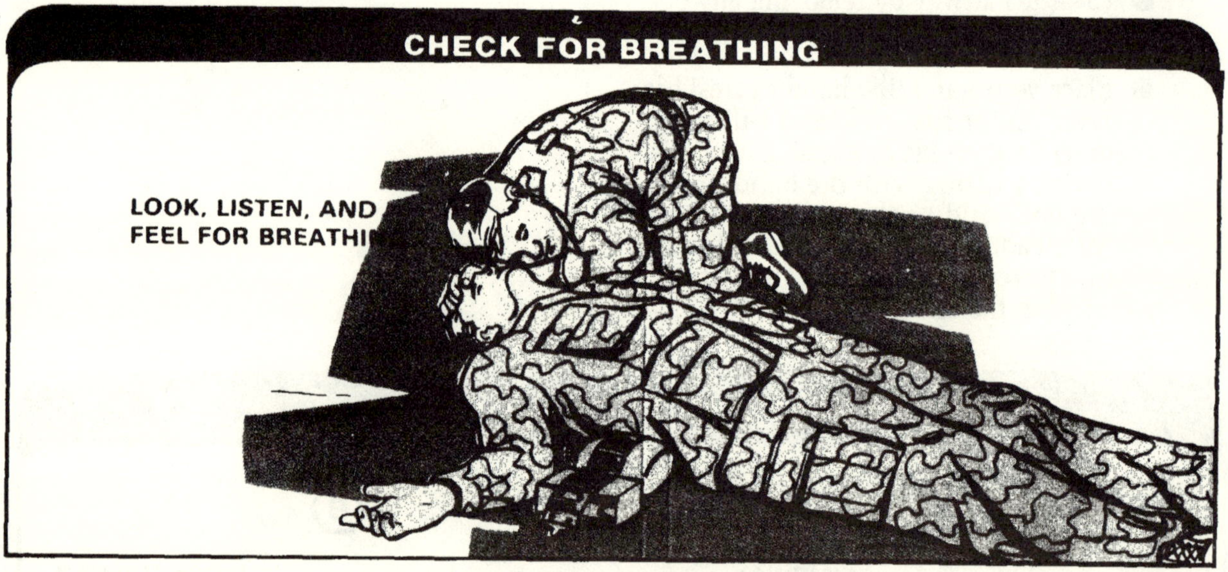

Restore Breathing. IF THERE ARE NO SIGNS OF BREATHING, START MOUTH-TO-MOUTH RESUSCITATION AT ONCE. The following procedures **should be used**

- **Put a hand under the casualty's neck to keep the head tilted far back.**
- **Press down on his forehead with the other hand.**
- **Move this hand and pinch his nostrils between your thumb and index finger.**
- **Open his mouth wide.**

- Take a deep breath and place your mouth over his, making an airtight seal with your lips.
- Blow into his mouth.
- Give four or five quick but full breaths to make sure his lungs are full.
- Remove your mouth, turn your head, and LOOK, LISTEN, and FEEL for exhaled air.
- Repeat this procedure once every 5 seconds until the casualty exhales.

GIVE MOUTH-TO-MOUTH RESUSCITATION

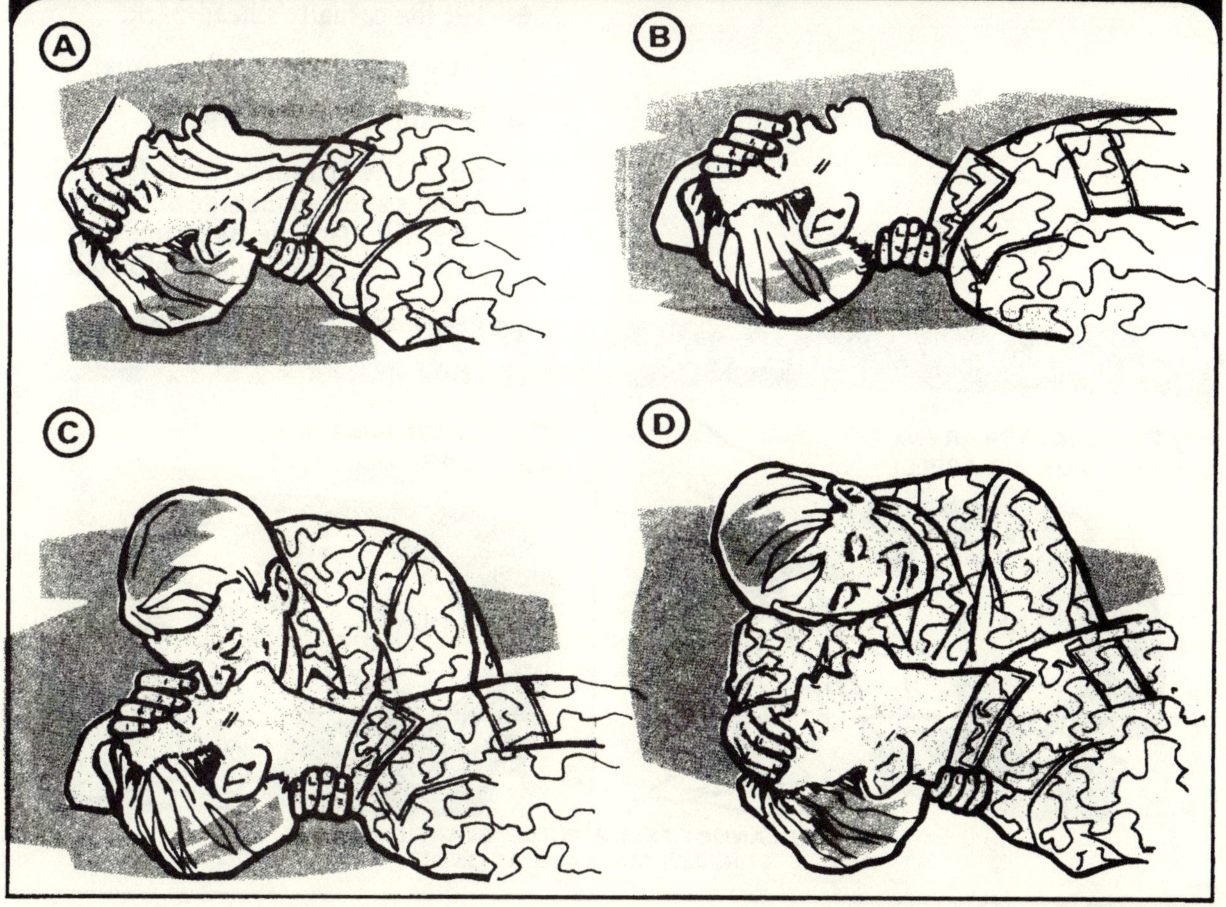

If you feel strong resistance when you first blow air into the casualty's mouth, quickly reposition his head and try again. If the airway is still not clear, roll him onto his side. Hit him sharply between his shoulder blades with the heel of your hand to dislodge any foreign objects. If the casualty's abdomen bulges (air going into stomach), apply gentle pressure on his abdomen with one hand to force the air out. If this makes the casualty vomit, quickly turn him onto his side, clean out his mouth, and continue giving mouth-to-mouth resuscitation.

Check for Heartbeat. When you find an unconscious casualty, check to see if he has a heartbeat and if he is breathing. To check for heartbeat, use the **following procedures:**

- **Tilt the casualty's head back.**
- **Place your fingers on his throat.**
- **Feel for the Adam's apple.**
- **Slide the fingers down from the Adam's apple to the side of the throat. This will place the fingertips over an artery, where the pulse can be felt.**

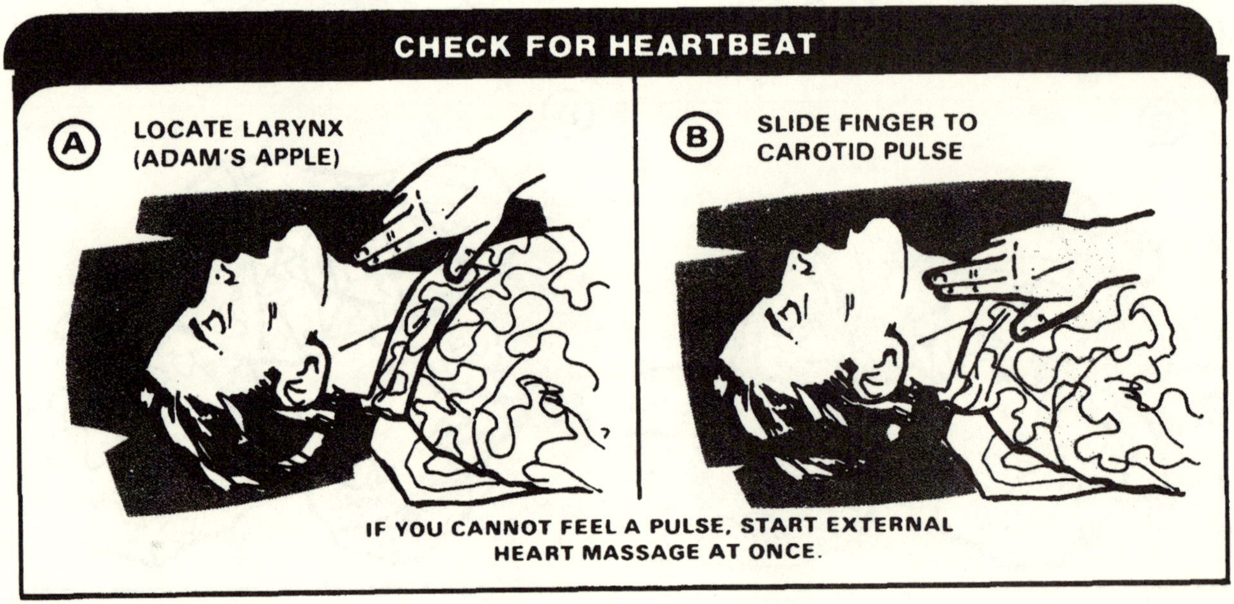

Restore Heartbeat. You must start external heart massage quickly, as permanent damage to the brain may occur if it is deprived of oxygenated blood. Examples of times without oxygen and likelihood of brain damage are **listed below:**

BRAIN DAMAGE WITHOUT OXYGEN
0-4 MIN BRAIN DAMAGE NOT LIKELY
4-6 MIN BRAIN DAMAGE PROBABLE
6-10 MIN BRAIN DAMAGE VERY LIKELY
OVER 10 MIN BRAIN DAMAGE ALMOST CERTAIN

External heart massage provides artificial circulation by squeezing the heart between the breastbone and the backbone, forcing blood through the lungs, brain, and body.

To perform mouth-to-mouth resuscitation and external heart massage at the **same time:**

- **Kneel at the casualty's side.**
- **Blow four quick but full breaths into the casualty (as described earlier) to fill the lungs with air (his head must be tilted back and his airway open). Locate the tip of the breastbone and measure two finger-widths up from that tip.**
- **Place the heel of the other hand along side the fingers. Then, put both hands together and interlace the fingers. Push downward on the chest 15 times at a rate of 80 counts per minute.**

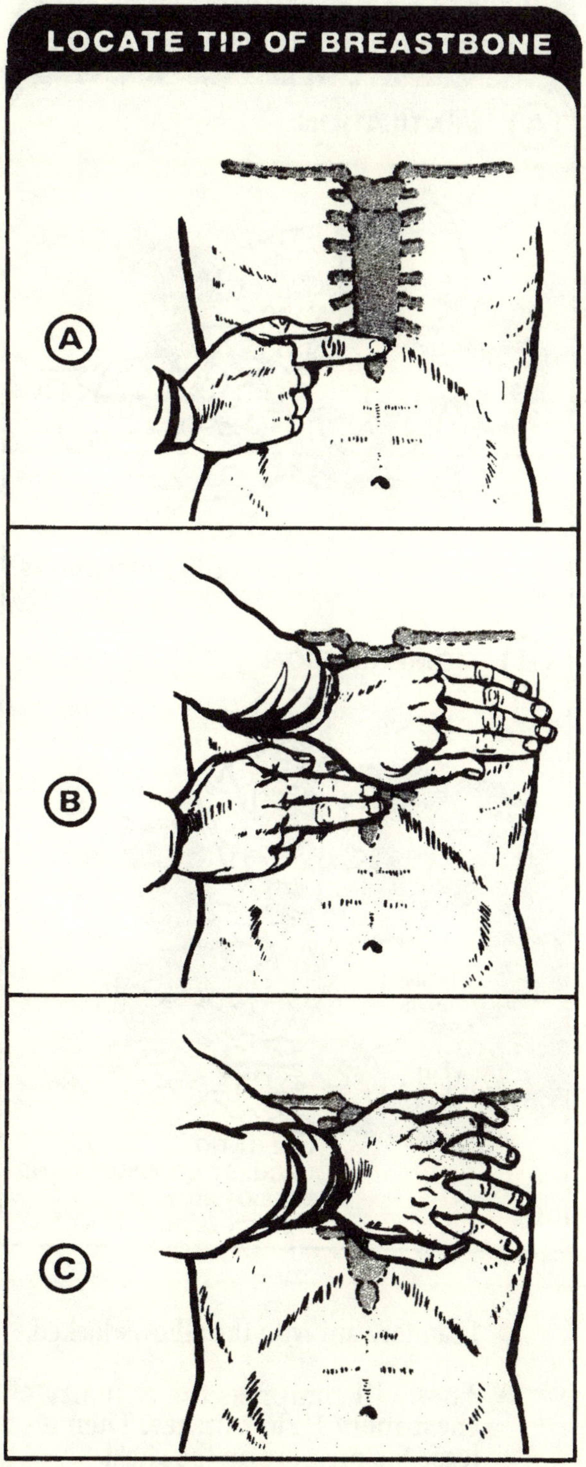

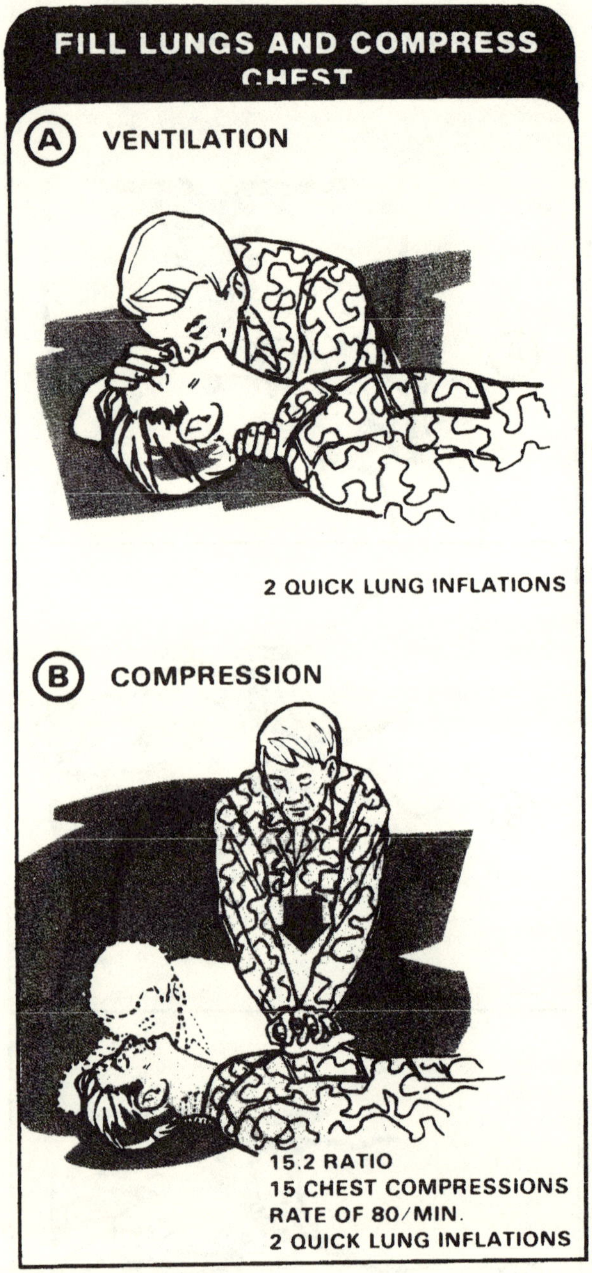

- Lean forward with the elbows locked.
- That will compress the casualty's chest about 1½ to 2 inches. Then release the pressure on the chest.

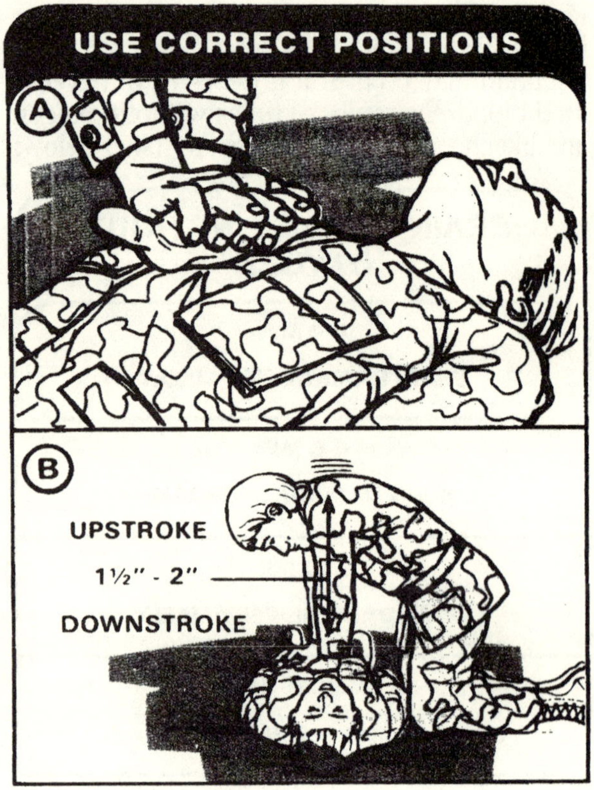

- **After each 15 compressions, shift positions slightly and give him 2 quick, but full, breaths.**
- **Continue this 15 to 2 ratio:**
  - ☐ **Until the casualty can breathe by himself and his pulse returns.**
  - ☐ **Until relieved by someone.**
  - ☐ **Until the casualty is dead.**

If two of you are present, one should give mouth-to-mouth resuscitation and the other should give heart massage. In that case, the procedure is slightly different. The soldier giving the heart massage should change the number of compressions from 15 at a time to 5, keeping the 80-per-minute rate. The soldier giving mouth-to-mouth resuscitation, should give 2 breaths after each 5 compressions.

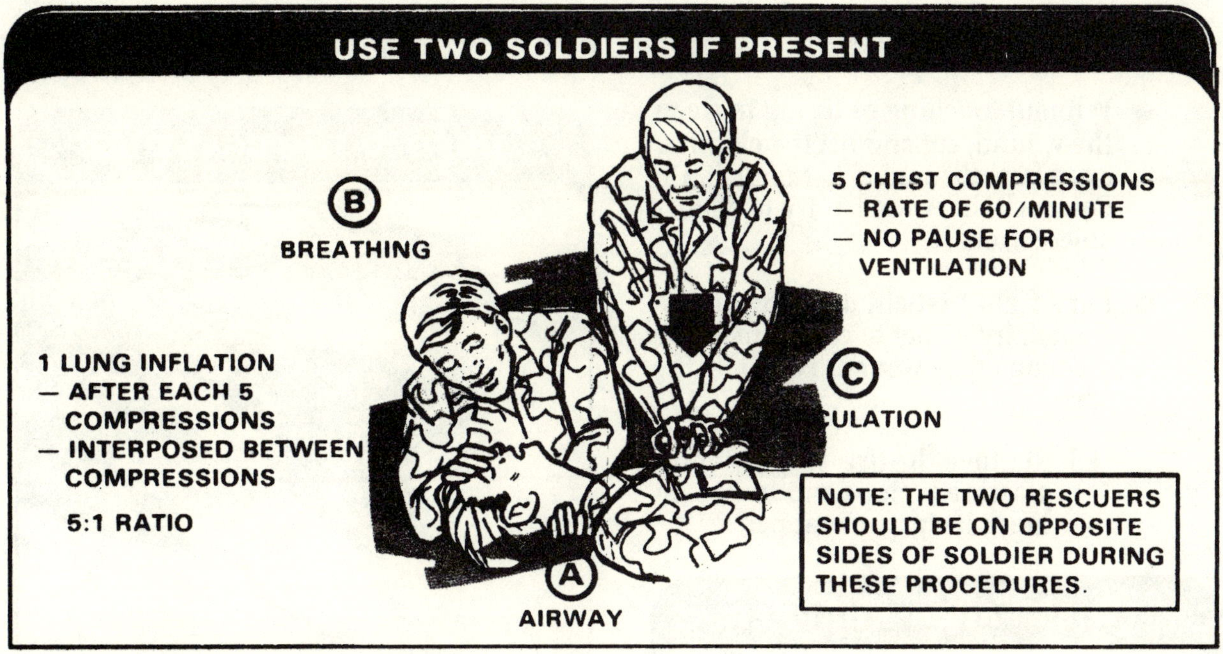

STOP THE BLEEDING

If the casualty is breathing and his heart is beating, the next thing to do is to stop the bleeding of the wound. Before you stop the bleeding, you must find all wounds. Look for both entry and exit points. This is to see that nothing is overlooked, as a bullet usually makes a smaller wound where it enters than where it exits.

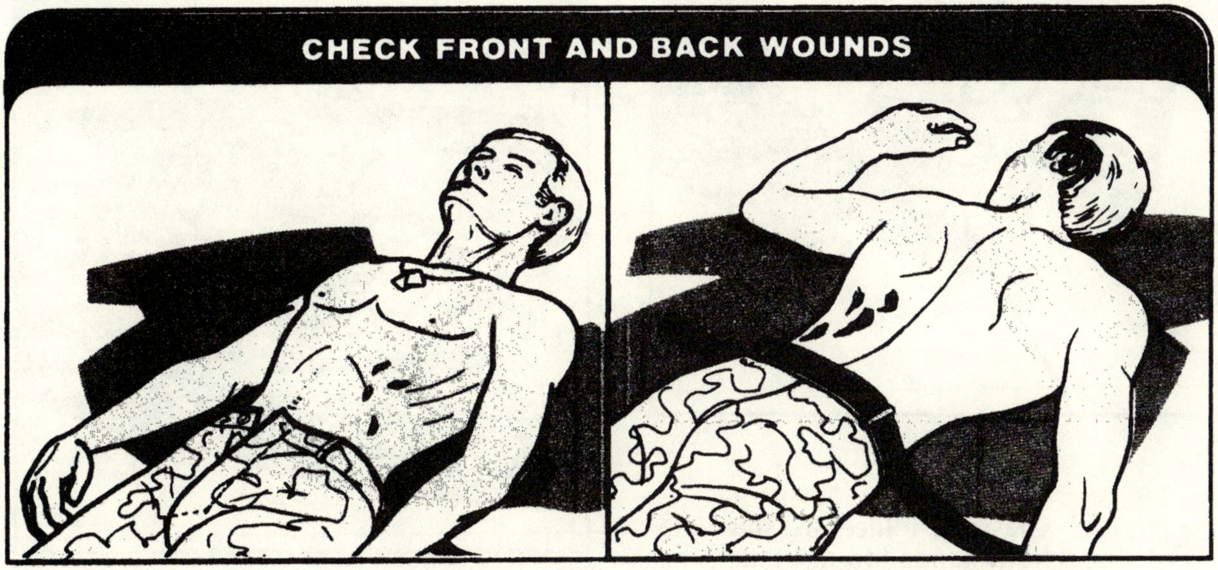

After finding all wounds, stop the bleeding by using the **following procedure:**

- **Without touching or trying to clean the wound, cut and lift the clothing away from the wound to expose it. Do not touch the wound or try to remove objects from it.**

- **Put a field first-aid dressing on the wound, trying not to contaminate the dressing or the wound. To put on the dressing:**

  ☐ **Remove the dressing from its plastic envelope and twist it to break the paper wrapper.**

  ☐ **Grasp the folded dressing with both hands (do not touch the side of the dressing that goes on the wound).**

  ☐ **Place the dressing on the wound without letting it touch anything else.**

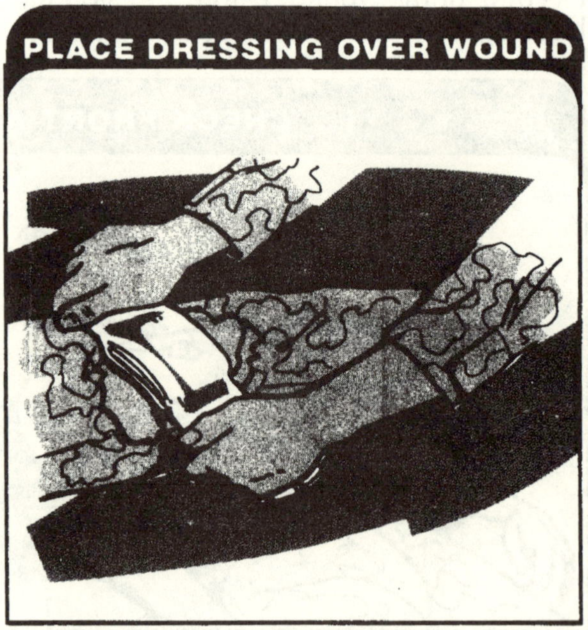

- ☐ Wrap the dressing around the wound and tie the ends securely with a square knot. If possible, tie the knot directly over the wound.

- If the bleeding continues after the dressing is secured on the wound, press the bandage for 5 to 10 minutes.

- If more pressure is needed to stop the bleeding, put a thick pad or stone on top of the dressing and tie the ends of the dressing over the pad or stone. This is called a pressure dressing.

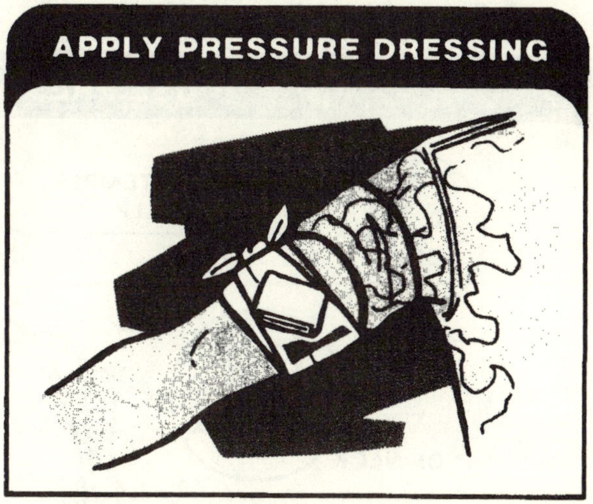

- If the wound is in an arm or leg and the bleeding has not stopped, raise the injured limb above the level of the heart. This helps to slow down or stop the bleeding. Do not, however, raise a limb with a broken bone unless it is properly splinted.

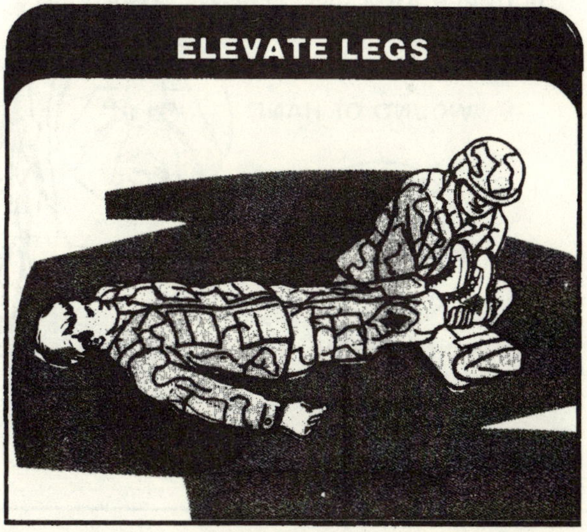

- If blood is spurting from the wound, there is bleeding from an artery. To stop it, press on the point of the body where the main artery supplying the wounded area with blood is located This pressure should shut off or slow down the flow of blood from the heart to the wound until a pressure dressing can be put on it. In some cases, you may have to keep pressure on the pressure paint even after the drawing is put on. The best pressure points of the body to use in stopping arterial bleeding are shown in the following illustration.

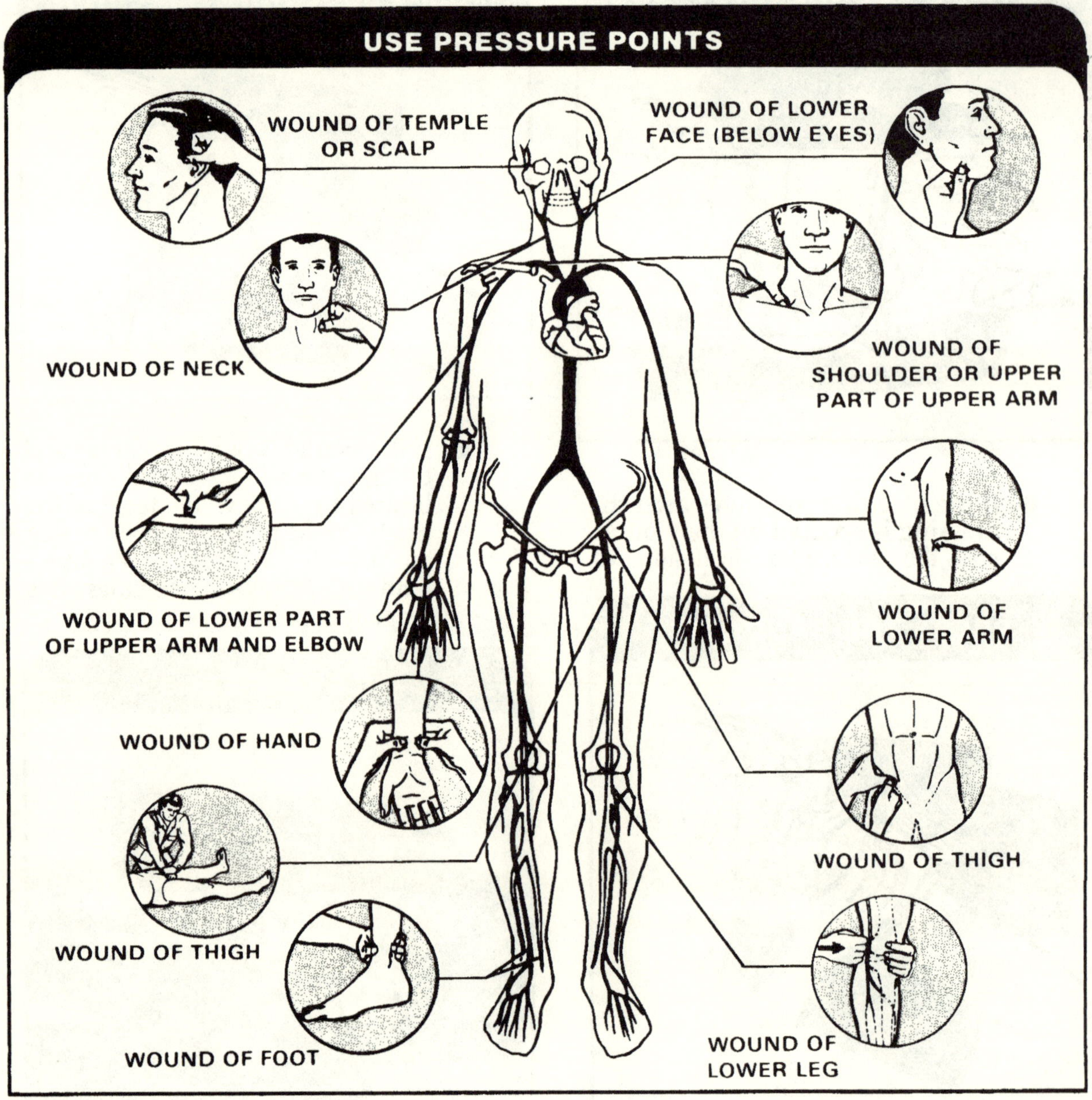

- If the wound continues to bleed after you apply pressure to a pressure point and apply a pressure dressing, use a tourniquet. This should be a LAST RESORT ONLY. Put the tourniquet between the wound and where the injured limb joins the trunk. Put it 2 to 4 inches above the wound, not over it. Never loosen or remove a tourniquet once it has been put on. If possible, mark a "T" on the casualty's forehead at the time the tourniquet is put on. Then get the casualty to an aid station quickly.

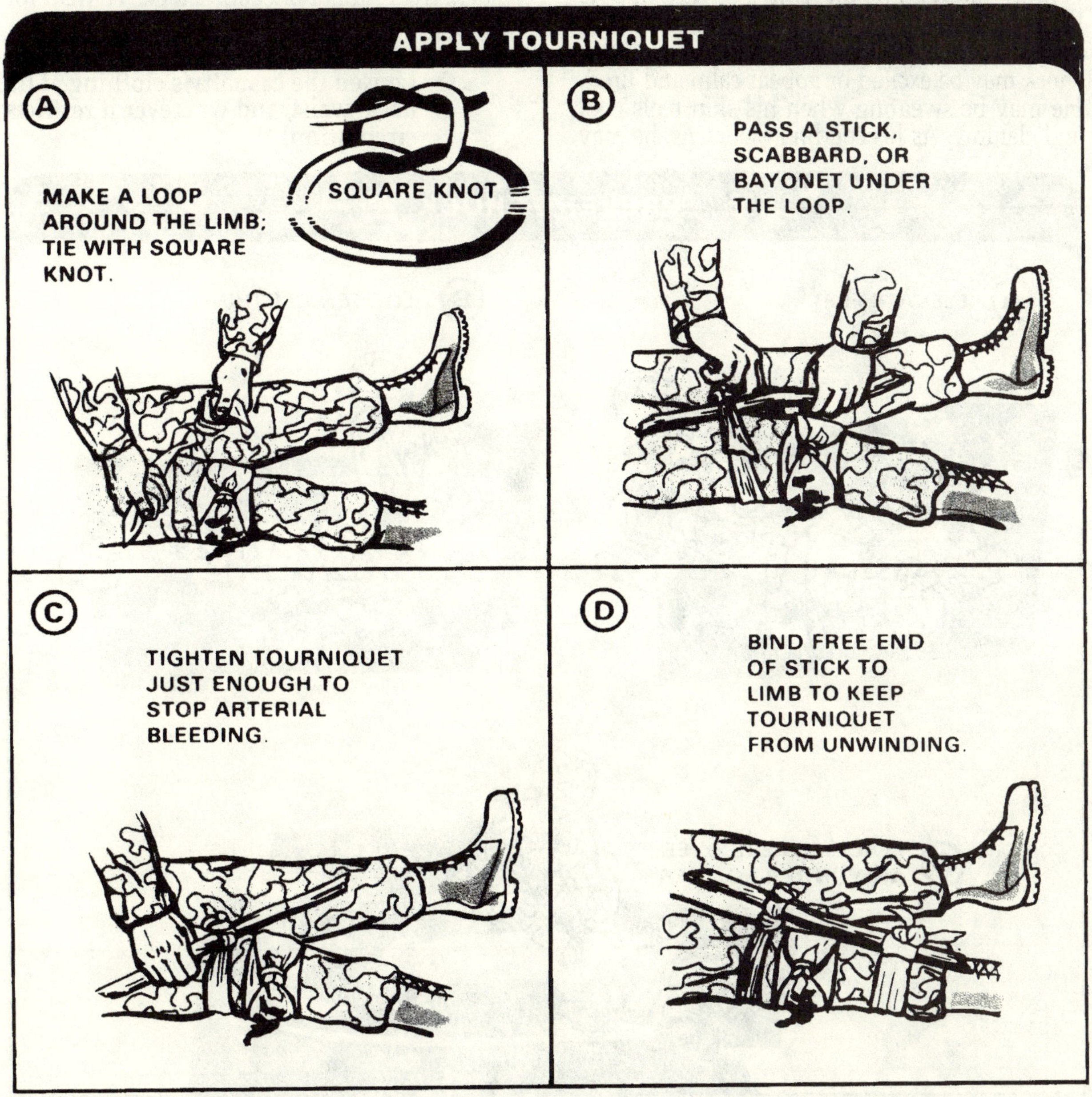

PREVENT SHOCK

Unless shock is prevented or treated, death may result, even though the injury would not otherwise be fatal.

Shock may result from any injury, but is more likely to result from a severe injury. **Warning signs of shock are restlessness, thirst, pale skin, and rapid heartbeat.** A casualty in shock may be excited or appear calm and tired. He may be sweating when his skin feels cool and clammy. As his condition worsens, he may take small, fast breaths or gasps; stare blindly into space or become blotchy or bluish around his mouth.

After giving the casualty the first two lifesaving measures, look for signs of shock. If the casualty is in shock or is about to go into shock, treat him at once for shock. To treat for shock, **proceed as follows:**

- **Loosen the casualty's clothing at the neck, waist, and wherever it restricts circulation.**

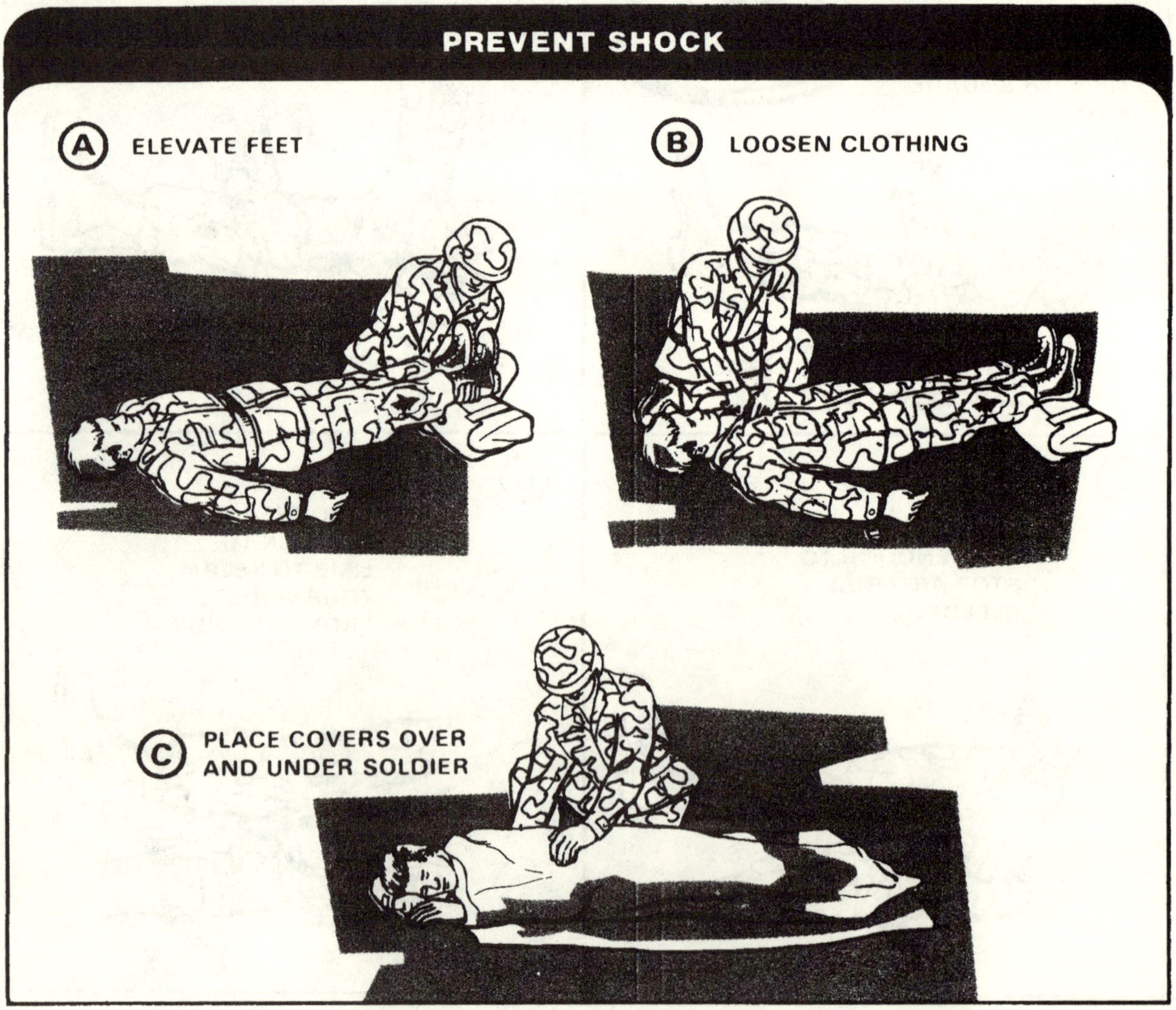

Reassure the casualty by being calm and self-confident. Assure him that he will be taken care of.

Place the casualty in a comfortable position. His position depends on his condition. If he is conscious, place him on his back with his feet raised 15 to 20 cm (6 to 8 in). If he is unconscious, place him on his side or abdomen with his head turned to the side. If he has a head wound, raise his head higher than his body. If he has a wound of the face and/or neck, set him up and lean him forward with his head down or in the position for an unconscious casualty. If he has a sucking chest wound, set him up or lay him down on the injured side. If he has an abdominal wound, lay him on his back with his head turned to the side.

Keep the casualty warm. It may be necessary to place ponchos or blankets under and over him.

DRESS AND BANDAGE THE WOUND

The healing of wounds and recovery depend a lot on how well you initially protect the wound from contamination and infection.

A wound must be dressed and bandaged to protect it from further contamination, as well as to stop the bleeding. Use the field first-aid dressing in the first-aid packet to dress' and bandage a wound. A dressing is any sterile pad used to cover a wound. A bandage is any material used to secure a dressing to a wound. The field first-aid dressing already has bandages attached to it. Use the dressing to cover a wound and the bandages to secure the dressing to it.

For information on how to dress and bandage different wounds, **see chapter 6, FM 21-11.**

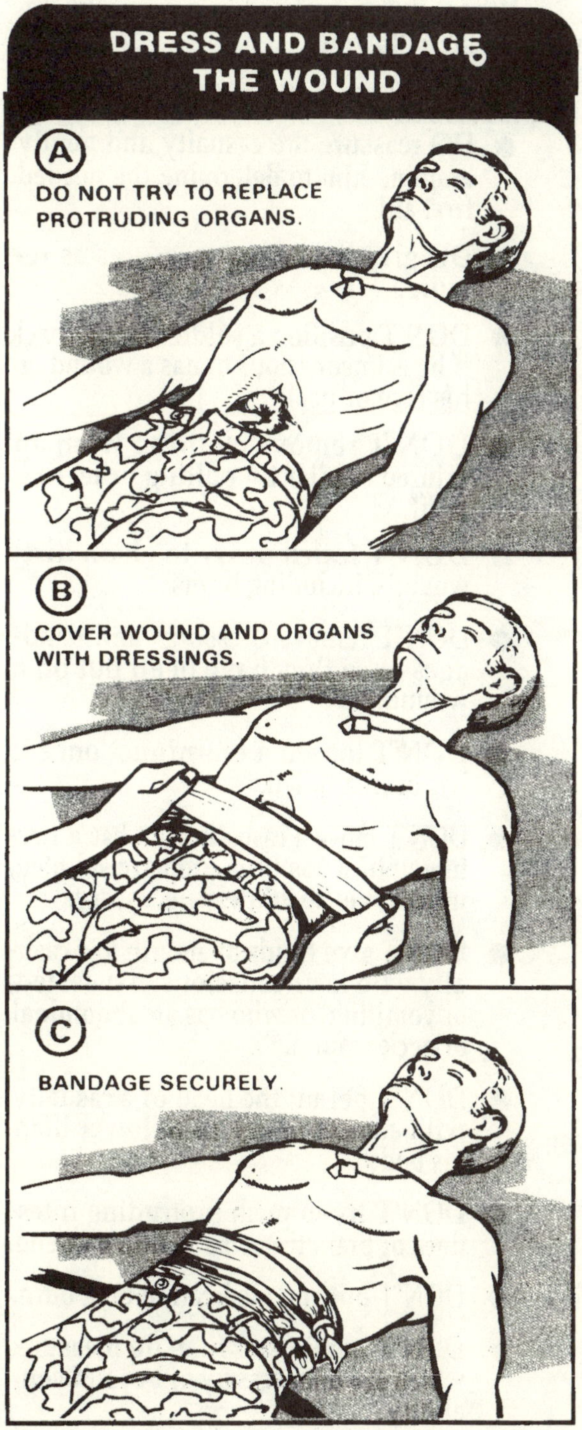

DOs AND DON'Ts OF FIRST AID

When giving first aid to a casualty, **remember the following:**

- **DO act promptly but calmly.**
- **DO reassure the casualty and gently examine him to determine the needed first aid.**
- **DO give lifesaving measures as required.**
- **DON'T position a soldier on his back if he is unconscious or has a wound on his face or neck.**
- **DON'T remove clothing from an injured soldier by pulling or tearing it off.**
- **DON'T touch or try to clean dirty wounds, including burns.**
- **DON'T remove dressings and bandages once they have been put on a wound.**
- **DON'T loosen a tourniquet once it has been applied.**
- **DON'T move a casualty who has a fracture until it has been properly splinted, unless it is absolutely necessary.**
- **DON'T give fluids by mouth to a casualty who is unconscious, nauseated, or vomiting, or who has an abdominal or neck wound.**
- **DON'T permit the head of a casualty with a head injury to be lower than his body.**
- **DON'T try to push protruding intestines or brain tissue back into a wound.**
- **DON'T put any medication on a burn.**
- **DON'T administer first-aid measures which are unnecessary or beyond your ability.**
- **DON'T fail to replace items used from the first-aid case.**

PERSONAL HYGIENE

Personal hygiene consists of practices which safeguard your health and that of others. It is often thought of as being the same as personal cleanliness. While cleanliness is important, it is only one part of healthy living. Personal hygiene is **important to you because:**

- **It protects against disease-causing germs that are present in all environments.**
- **It keeps disease-causing germs from spreading.**
- **It promotes health among soldiers.**
- **It improves morale.**

PERSONAL CLEANLINESS

Skin. Wash your body frequently from head to foot with soap and water. If no tub or shower is available, wash with a cloth and soapy water, paying particular attention to armpits, groin area, face, ears, hands, and feet.

Hair. Keep your hair clean, neatly combed, and trimmed. At least once a week, wash your hair and entire scalp with soap and water. Also, shave as often as the water supply and tactical situation permit. Do not share combs or shaving equipment with other soldiers.

Hands. Wash your hands with soap and water after any dirty work, after each visit to the latrine, and before eating. Keep your fingernails closely trimmed and clean. Do not bite your fingernails, pick your nose, or scratch your body.

Clothing and Sleeping Gear. Wash or exchange clothing when it becomes dirty (situation permitting). Wash or exchange sleeping gear when it becomes dirty. If clothing and sleeping gear cannot be washed or exchanged, shake them and air them regularly in the sun. That greatly reduces the number of germs on them.

CARE OF THE MOUTH AND TEETH

Regular and proper cleaning of the mouth and the teeth helps prevent tooth decay and gum disease. The most healthful oral hygiene is to clean your mouth and teeth thoroughly and correctly after each meal with a toothbrush and toothpaste. If a toothbrush is not available, cut a twig from a tree and fray it on one end to serve as a toothbrush. If mouthwash is available, use it to help kill germs in your mouth. To help remove food from between your teeth, use dental floss or toothpicks. Twigs can also be used for toothpicks.

MAKE A TOOTHBRUSH

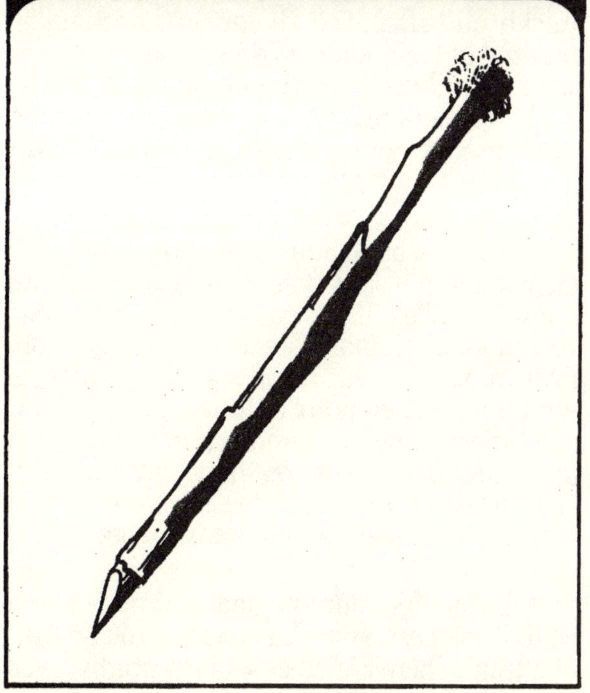

Wash and dry your feet daily. Use foot powder on your feet to help kill germs, reduce friction on the skin, and absorb perspiration. Socks should be changed daily. After crossing a wet area, dry your feet, put on foot powder, and change socks, as soon as the situation permits.

CHANGE SOCKS AND POWDER FEET

FOOD AND DRINK

For proper development, strength, and survival, **your body requires:**

- **Proteins.**
- **Fats and carbohydrates.**
- **Minerals.**
- **Vitamins.**
- **Water.**

Issued rations have those essential food substances in the right amounts and proper balance. So, eat primarily those rations. When feasible, heat your meals. That will make them taste better and will reduce the energy required to digest them. Do not overindulge in sweets, soft drinks, alcoholic beverages, and other non-issued rations. Those rarely have nutritional value and are often harmful.

Drink water only from approved water sources or after it has been treated with water-purification tablets. To purify water from **rivers or streams:**

- **Fill your canteen with water (be careful not to get trash or other objects in your canteen).**
- **Add one purification tablet per quart of clear water or two tablets per quart of cloudy or very cold water. (If you are out of tablets, use boiling water that has been boiled for 5 minutes.)**
- **Replace the cap loosely.**
- **Wait 5 minutes.**
- **Shake the canteen well and allow some of the water to leak out.**
- **Tighten the cap.**
- **Wait an additional 20 minutes before drinking the water.**

EXERCISE

Exercise of the muscles and joints helps to maintain physical fitness and good health. Without that, you may lack the physical stamina and ability to fight. Physical fitness includes a healthy body, the capacity for skillful and sustained performance, the ability to recover from exertion rapidly, the desire to complete a designated task, and the confidence to face any eventuality. Your own safety, health, and life may depend on your physical fitness.

There are lulls in combat when you will not be active. During such lulls, exercise. That helps to keep the muscles and body functions ready for the next period of combat. It also helps pass the time in the lulls.

REST

Your body needs regular periods of rest to restore physical and mental vigor. When you are tired, your body functions are sluggish, and your ability to react is slower than normal. That also makes you more susceptible to sickness. For good health, 6 to 8 hours of uninterrupted sleep each day is desirable. As that is seldom possible in combat, use rest periods and off-duty time to rest or sleep. Do not be ashamed to say that you are tired or sleepy. Do not, however, sleep when on duty.

MENTAL HYGIENE

The way you think affects the way you act. If you know your job, you will probably act quickly and effectively. If you are uncertain or doubtful of your ability to do your job, you may hesitate and make wrong decisions. Positive thinking is a necessity. You must enter combat with absolute confidence in your ability to do your job.

Fear is a basic human emotion. It is both a mental and physical state. Fear is not shameful if it is controlled. It can even help you by making you more alert and more able to do your job. Fear makes the pupils of your eyes enlarge, which increases your field of vision so you can detect movement more easily. Fear also increases your rate of breathing and heartbeat. That increases your strength. Therefore, control your fear and use it to your advantage.

Do not let your imagination and fear run wild. Remember, you are not alone. You are part of a team. There are other soldiers nearby, even though they cannot always be seen. Everyone must help each other and depend on each other.

Worry undermines the body, dulls the mind, and slows down thinking and learning. It adds to confusion, magnifies troubles, and causes you to imagine things which really do not exist. If you are worried about something, talk to your leader about it. He may be able to help solve the problem.

You may have to fight in any part of the world and in all types of terrain. Therefore, adjust your mind to accept conditions as they are. If mentally prepared for it, you should be able to fight under almost any conditions.

RULES FOR AVOIDING ILLNESS IN THE FIELD

- **Don't consume foods and beverages from unauthorized sources.**
- **Don't soil the ground with urine or feces. (Use a latrine or "cat-hole.")**
- **Keep your fingers and contaminated objects out of your mouth.**
- **Wash your hands following any contamination, before eating or preparing food, and before cleaning your mouth and teeth.**
- **Wash all mess gear after each meal.**
- **Clean your mouth and teeth at least once each day.**
- **Avoid insect bites by wearing proper clothing and using insect repellents.**
- **Avoid getting wet or chilled unnecessarily.**
- **Don't share personal items (canteens, pipes, toothbrushes, washcloths, towels, and shaving gear) with other soldiers.**
- **Don't leave food scraps lying around.**
- **Sleep when possible.**
- **Exercise regularly.**

APPENDIX A

Mines

GENERAL

A unit may use mines during security, defensive, retrograde, and offensive operations in order to reduce the enemy's mobility. In those operations, leaders pick the places for the mines and their men emplace them and, when required, retrieve them (See TM 9–1345-203-12P).

The mines you will most commonly use are:

- **M14, Antipersonnel**
- **M16Al, Antipersonnel**
- **M18Al, Antipersonnel**
- **M26, Antipersonnel**
- **M15, Antitank**
- **M21, Antitank**
- **M24, Off-Route Antitank**

CONTENTS

GENERAL A-1
ANTIPERSONNEL A-2
ANTITANK A-22

ANTIPERSONNEL

M14, ANTIPERSONNEL MINE

This is a blast-type, high-explosive mine with a plastic body. A pressure of 9 to 15.8kg (20 to 35 lb) will detonate it.

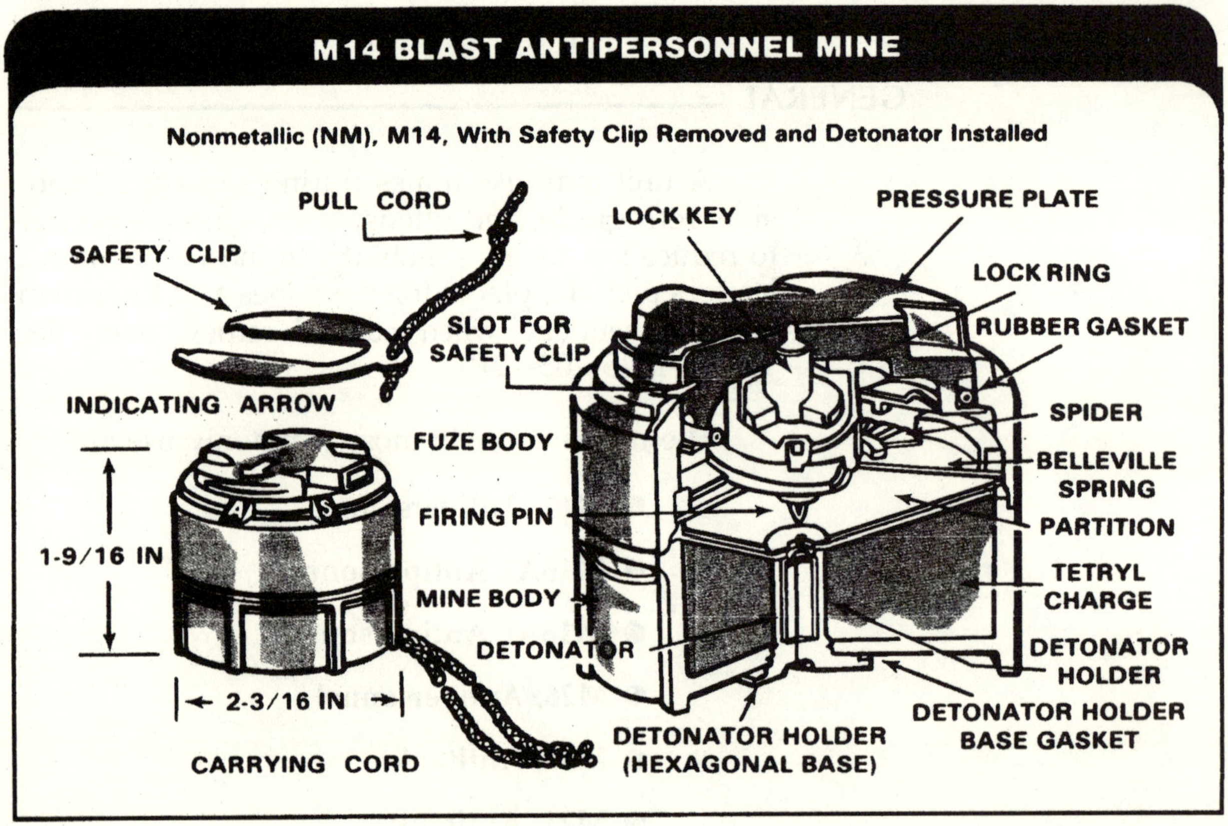

To emplace an **M14 mine:**

- **Remove the mine from the packing box and inspect it. If the mine is cracked or otherwise damaged, do not use it.**
- **Use the M22 wrench from the packing box to unscrew the white plastic shipping plug from the detonator well in the bottom of the mine. Keep the shipping plug for possible future use.**

FUZE WRENCH M22

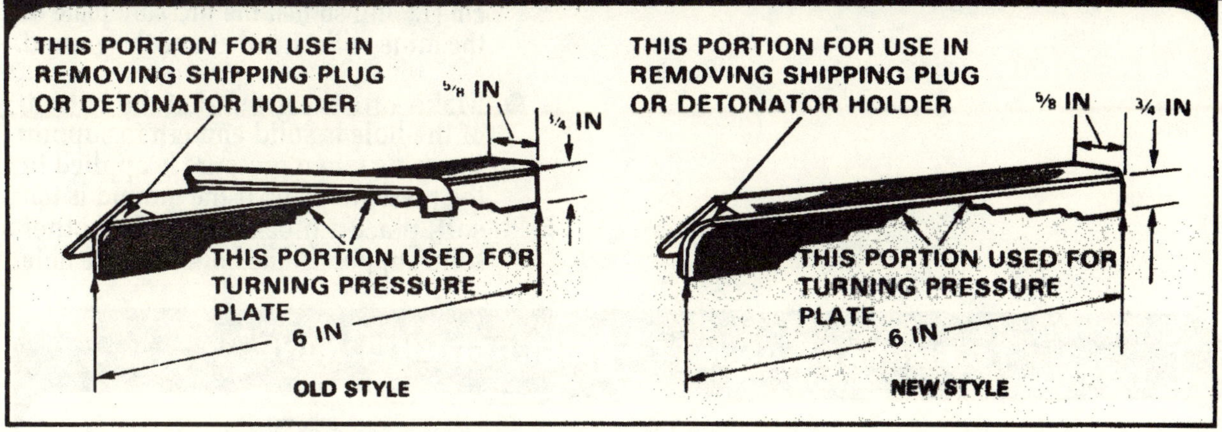

- Inspect the firing pin's position. If it extends into the detonator well. the mine is unsafe to use.

INSPECTING PIN'S POSITION

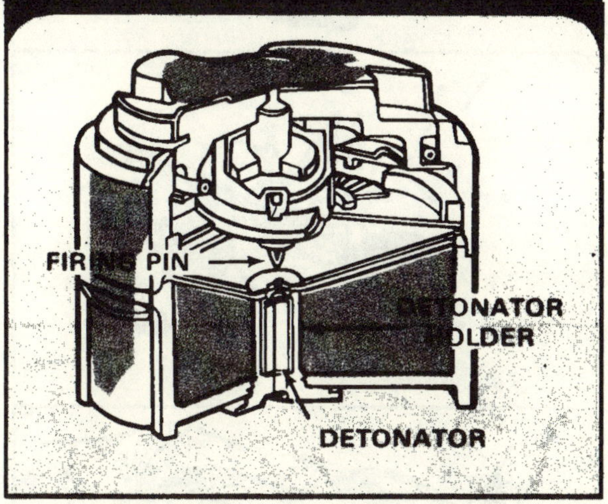

- Inspect the detonator well for foreign matter. If foreign matter is present, carefully remove it by tapping the mine against the palm of the hand.

- Dig a hole about 10 cm (4 in) in diameter and just deep enough (about 3.8 em [1.5 in]) so that the pressure plate of the mine will extend above the ground.
- Make sure the ground at the bottom of the hole is solid enough to support the mine when pressure is applied to the pressure plate. If the ground is too soft, place a block of wood or other solid support in the bottom of the hole.

TO EMPLACE AN M14 ANTIPERSONNEL MINE

GRASP THE MINE IN ONE HAND AND TURN THE MINE SO THAT THE FUZE WITH THE SAFETY CLIP IS FACE UP. WITH YOUR OTHER HAND, PULL ON THE CARRING CORD ATTACHED TO THE SAFETY CLIP. KEEP THE SAFETY CLIP FOR DISARMING THE MINE LATER, IF REQUIRED.

UNSCREW SHIPPING PLUG FROM BOTTOM OF MINE. TURN PRESSURE PLATE TO ARMED POSITION WITH ARMING TOOL.

REMOVE SAFETY CLIP AND CHECK FOR MALFUNCTIONING.

TO EMPLACE AN M14 ANTIPERSONNEL MINE (CONTINUED)

REPLACE SAFETY CLIP.

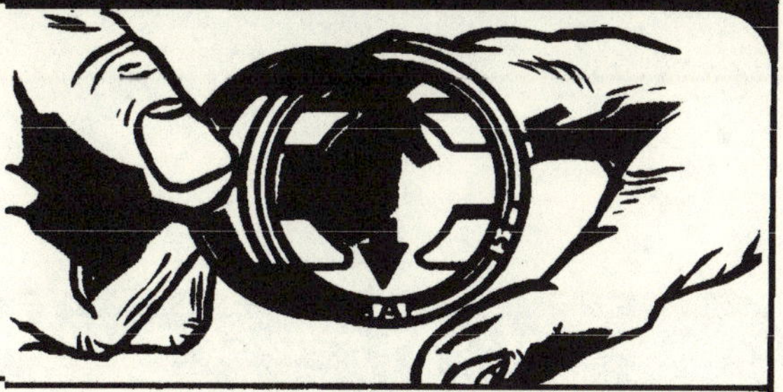

SCREW DETONATOR INTO DETONATOR WELL.

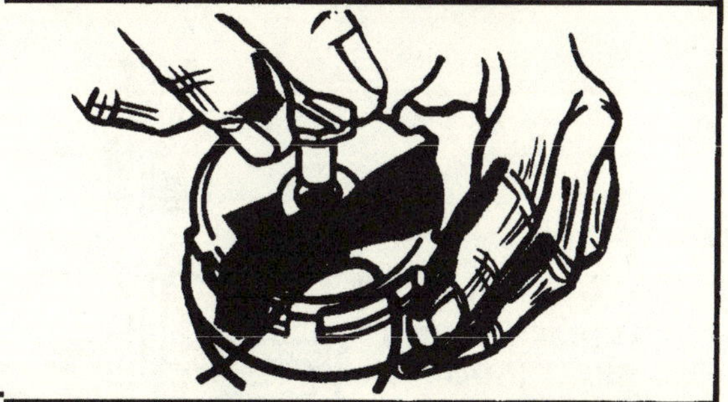

TO BURY: PRESSURE PLATE SHOULD BE SLIGHTLY ABOVE GROUND LEVEL.

BURY MINE AND REMOVE SAFETY CLIP.

USE THE M22 WRENCH TO ARM THE MINE BY TURNING THE PRESSURE PLATE CLOCKWISE FROM S TO A (SAFE TO ARMED(. IF THE PRESSURE PLATE SNAPS DOWNWARD SO THAT THE BODY OF THE MINE, AND THE SAFETY CLIP CANNOT BE INSERTED, DO NOT USE THE MINE.

To disarm and remove an M14 mine, reverse the steps used to arm and emplace it.

- Inspect the area around the mine to see if the mine has been tampered with. If it has been, do not try to disarm it. Report the tampering to your leader.
- Remove the soil from the mine without putting pressure on the mine.
- Grasp the body of the mine with one hand and insert the safety clip with the other.
- With the safety clip in place, turn the pressure plate so that the arrow points to S (SAFE). That disarms the mine.
- Remove the mine from the hole.
- Turn the mine over and carefully remove the detonator from the detonator well.
- Screw the plastic shipping plug into the detonator well.
- Clean off the mine and put it in a packing box.

M16Al, ANTIPERSONNEL MINE

This is a bounding, fragmentation mine with a metallic body. It can be set for pressure detonation or set with a tripwire attached to a release-pin ring. A pressure of 3.6 kg (8 lb) or more against one or more of the three prongs of the fuze, or a pull of 1.3 kg (3 lb) or more on the tripwire, will detonate the mine.

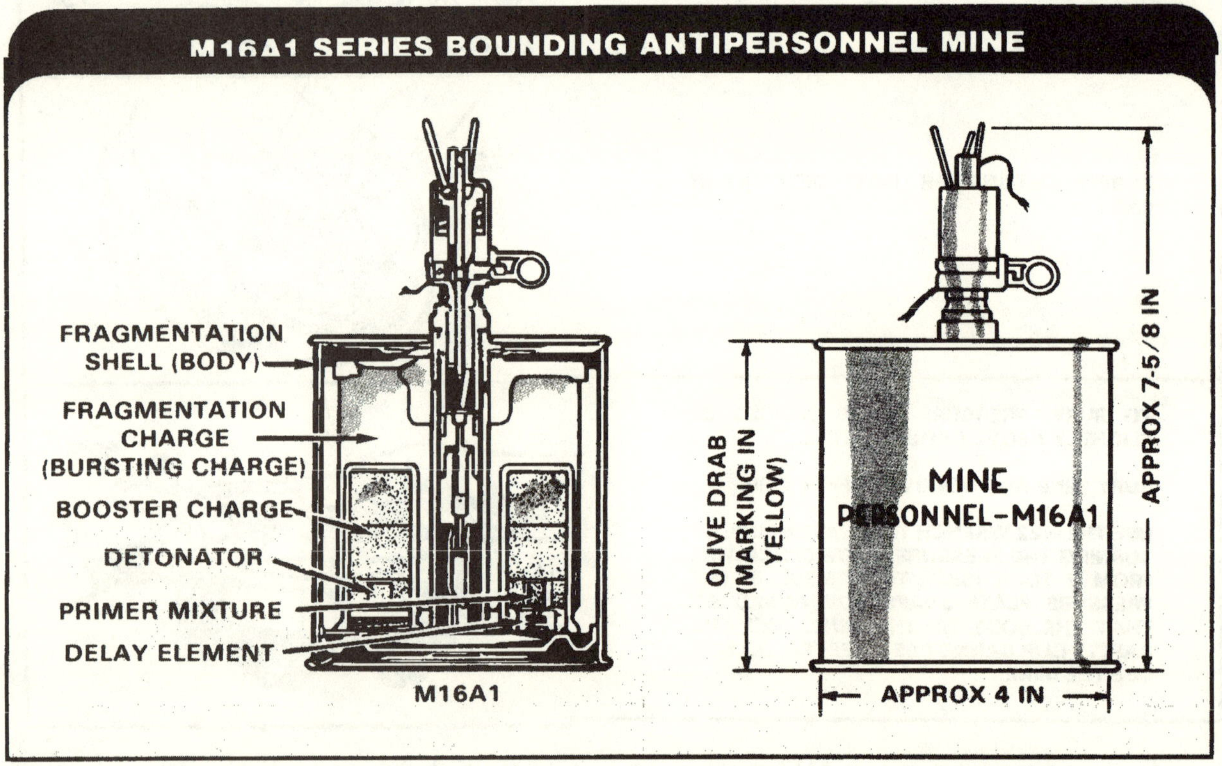

To emplace an **M16Al mine:**

- **Remove the mine from its packing box and inspect it for damage. If the mine is dented, cracked, or otherwise damaged, do not use it.**
- **Unscrew the shipping plug from the fuze well with the closed end of the M25 fuze wrench. Keep the shipping plug for future disarming of the mine, if required.**

FUZE WRENCH M25

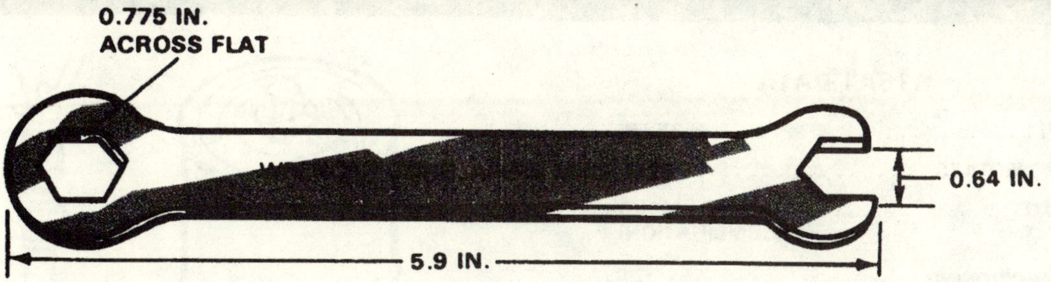

EXAMINING M16A1 MINE

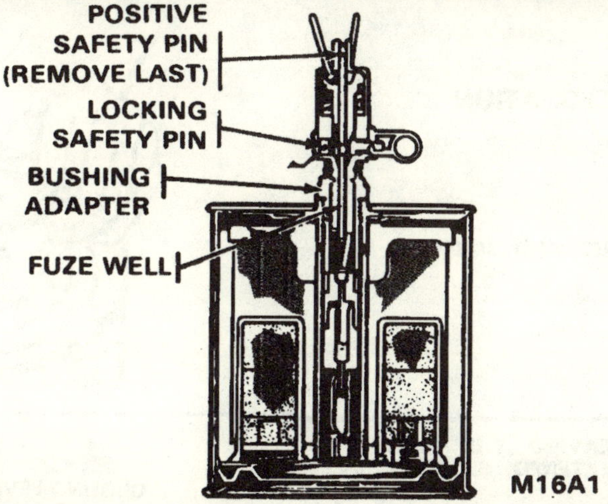

- Inspect the fuse well and flash tube of the mine for foreign matter. If foreign matter is present, turn the mine upside down and gently tap its bottom to dislodge the matter.
- Set the mine down and take a fuze out of the fuze box.
- Inspect the fuze for damage and for missing safety pins. Make sure that the safety pins move freely in the safety-pin holes. Also make sure that the rubber gasket is around the fuse base.
- With the open end of the wrench, make sure that the bushing adapter on the fuze well is tight.
- Screw the fuze assembly into the fuze well with the fuze wrench.
- Dig a hold about 15 cm (6 in) deep and 13 cm (5 in) in diameter.
- Put the mine in the hole.

TO EMPLACE AN M16A1 ANTIPERSONNEL MINE

M16A1 DATA

WT . . . . 8.25 LB.

PROJECTILES . . . . STEEL

FUZE . . . . M605 (COMBINATION)

FUNCTIONING
PRESSURE . . . . 8 to 20 LBS
PULL . . . . 3 to 10 LBS

BOUNDING HT 0.6–1.2M

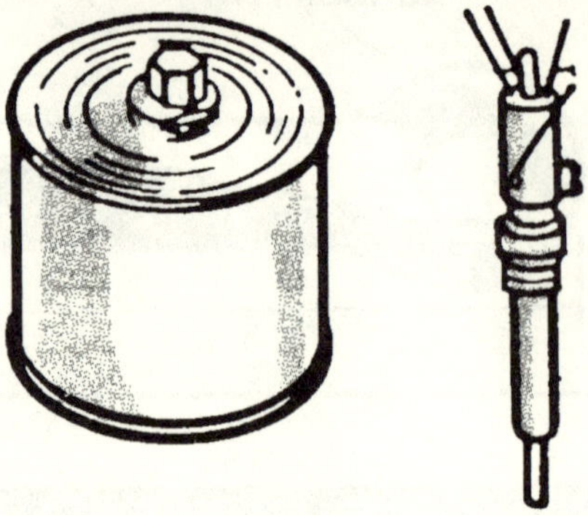

PRESSURE DETONATION

REMOVE SHIPPING PLUG AND SCREW IN FUZE

BURY THE MINE, LEAVING THE FUZE PRESSURE PRONGS EXTENED SLIGHTLY ABOVE GROUND LEVEL.

COVER THE MINE WITH DIRT, PRESSING IT FIRMLY AROUND THE SIDES OF THE MINE. LEAVE THE HEAD OF THE FUZE EXPOSED.

REMOVE THE LOCKING SAFETY PIN, THEN REMOVE THE INTERLOCKING SAFETY PIN FROM THE POSITIVE SAFETY PIN. KEEP THE SAFETY PINS FOR FUTURE DISARMING, IF REQUIRED.

ARRANGE THE PULL CORD ON THE POSITIVE SAFETY PIN SO THAT IT WILL PULL OUT EASILY.

COMOUFLAGE THE MINE.

REMOVE THE POSITIVE SAFETY PIN TO ARM THE MINE. IF THE PIN IS HARD TO REMOVE, DO NOT FORCE IT. GET ANOTHER FUZE ASSEMBLY AND START OVER.

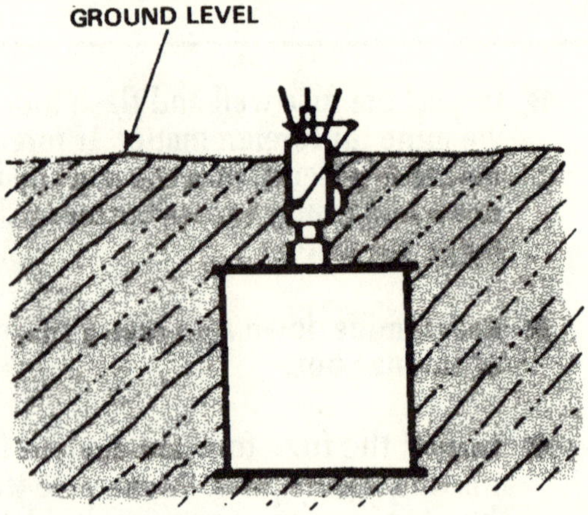

TO EMPLACE AN M16A1 ANTIPERSONNEL MINE (CONTINUED)

TRIP WIRE DETONATION

COVER THE MINE WITH SOIL, LEAVING THE RELEASE PIN RING OF THE FUZE AND PRESSURE PRONGS EXPOSED.

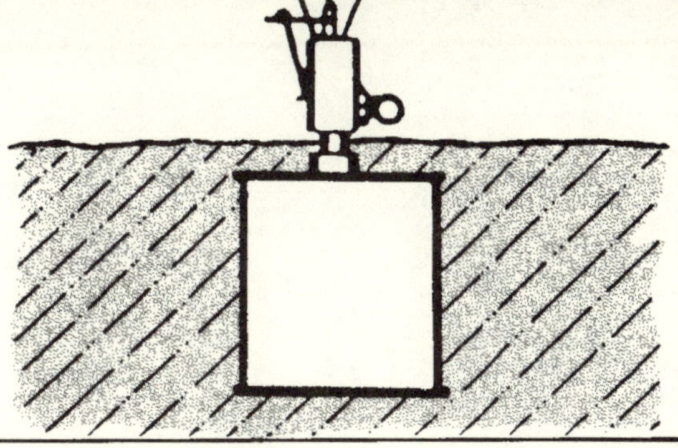

CASUALTY RADIUS 30M

DRIVE TWO ANCHOR STAKES INTO THE GROUND ABOUT 10 METERS (33 FT) FROM THE MINE SO THAT THE TRIPWIRES FROM A WIDE "V" WHEN ATTACHED TO THE MINE AND STAKES.

ATTACH A SEPARATE WIRE SECURELY TO EACH STAKE AND TO THE RELEASE PIN RING. LEAVE SLACK IN THE WIRE SO THAT NO PULL WILL BE EXERTED ON THE RELEASE PIN RING WHEN THE SAFETY PINS ARE REMOVED.

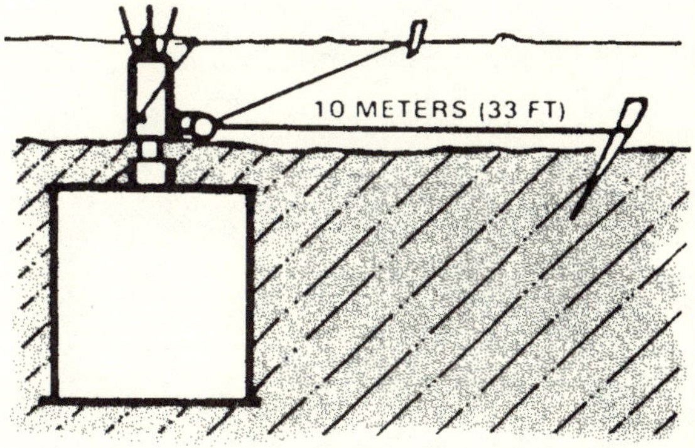

REMOVE THE LOCKING SAFETY PIN, THE THE INTERLOCKING PIN FROM THE POSITIVE SAFETY PIN. KEEP THE SAFETY PINS FOR FUTURE DISARMING, IF REQUIRED.

ARANGE THE PULL CORD ON THE POSITIVE SAFETY PIN SO THAT IT WILL PULL OUT EASILY.

CAMOUFLAGE THE MINE.

REMOVE THE POSITIVE SAFETY PIN TO ARM THE MINE. IF THE PIN IS HARD TO REMOVE, DO NOT FORCE IT. GET ANOTHER FUZE ASSEMBLY AND START OVER.

Mine bounds into air and explodes at height of 0.6 meters to 1.2 meters.

To disarm and remove an M16A1 mine, reverse the steps used to arm and emplace it.

- **Check the mine and the area around it to see if the mine has been tampered with.**
- **If it has been, do not try to disarm it.**
- **Report the tampering to your leader.**
- **Uncover the top of the mine.**
- **Insert the positive safety pin through the positive safety pin hole.**
- **Insert the locking safety pin through the locking safety pin hole opposite the release-pin ring.**
- **Insert the interlocking safety pin between the positive safety pin and locking safety pin.**
- **If tripwires are attached to the release-pin ring, cut all of them after the safety pins have been inserted.**
- **Remove the dirt from around the mine and then lift the mine out of the hole.**
- **Unscrew and remove the fuze assembly.**
- **Replace the plastic shipping plug in the fuze well.**
- **Replace the mine and fuze in the packing box.**

M18Al, ANTIPERSONNEL MINE (CLAYMORE)

This is a curved, rectangular mine containing C4 explosive and 700 steel balls. It can be fired electrically or nonelectrically.

The Claymore projects 700 steel balls in a fan-shaped pattern about 2 meters (6.6 ft) high and 60 degrees wide to a range of 50 meters (165 ft). These balls are effective as far as 100 meters (328 ft) and are dangerous up to 250 meters (825 ft) forward of the mine.

M18A1 FRAGMENTATION ANTIPERSONNEL MINE

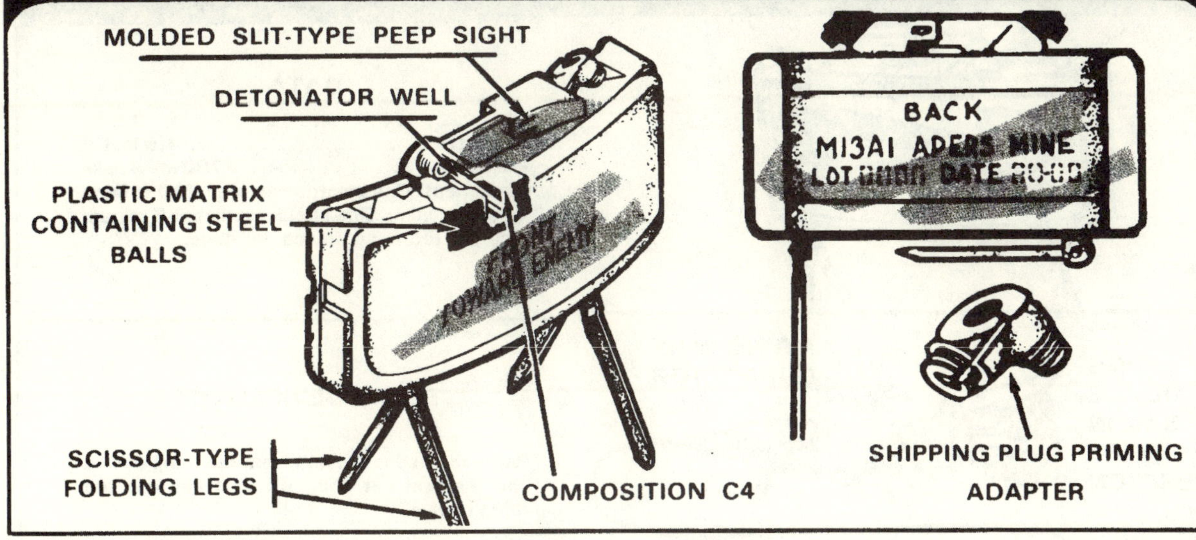

To emplace a Claymore for **command detonation:**

- Remove the electrical firing wire, the firing device, and test set from the bandoleer. Do not take the mine out of the bandoleer.
- Position the firing-device safety bail in the FIRE position and squeeze the firing-device handle with a firm, quick squeeze.

ANTIPERSONNEL MINE M18A1 AND ACCESSORIES PACKED IN BANDOLEER M7

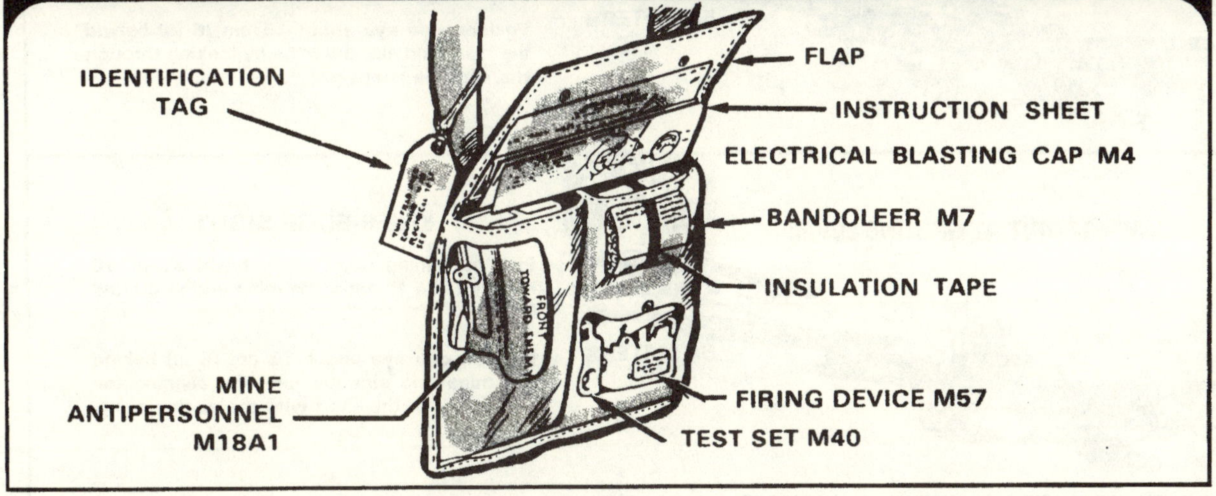

TO EMPLACE AN M18A1 ANTIPERSONNEL MINE

DATA

Wt. . . . . 3.5 lb.
Explosive . . . . 1.5 lb. C4
Projectiles . . . . 700 steel balls
Equipment: One electric cap with 100 ft firing wires per mine. One circuit tester per 6 mines. One electirc firing device per mine.

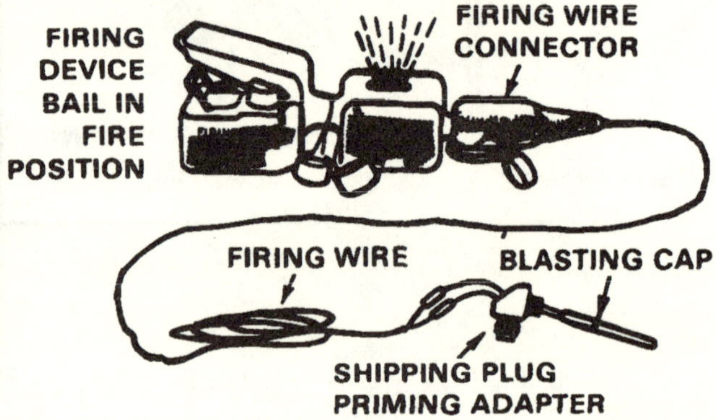

FIRING DEVICE TEST

Remove the dust covers from the firing device and the test set. Plug the test set into the firing device.

Position the firing device bail at the FIRE position, squeeze the handle on the firing device, and watch the window of the test set for a flash of light. The light means that the firing device and test set are functioning properly.

SLIT-TYPE PEEPSIGHT

Pick an aiming point (tree, bush) about 50 meters (164 ft) from the mine and about 2.5 meters (8 ft) above the ground.

Position one eye about 15 cm (6 in) behind the mine and aim the mine by looking through the slit-type peepsight.

KNIFE-EDGE SIGHT

Pick an aiming point (tree, bush) about 50 meters (164 ft) from the mine and at ground level.

Position one eye about 15 cm (6 in) behind the mine and aim the mine by aligning the two edges of the sight with the aiming point.

TO EMPLACE AN M18A1 ANTIPERSONNEL MINE (CONTINUED)

PREPARE TO FIRE

Unscrew one of the shipping plug priming adapters and keep if for possible later use.

Slide the slotted end of the shipping-plug priming adapter onto the firing wire of the blasting cap between the crimped connections and the blasting cap.

Insert the blasting cap into the detonator well and acrew on the adapter.

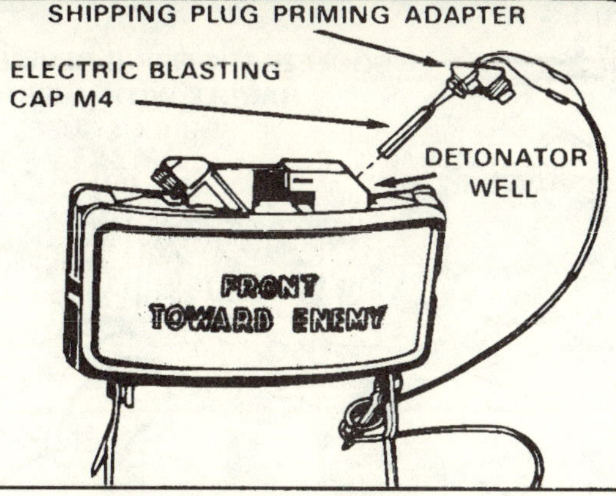

POSITION MINE

Tie the shorting-plug end of the firing wire to a fixed object (stake, tree) at the firing position.

Unroll firing wire and connect directly to firing device with the safety engaged.

Position the mine on the ground with the surface marked FRONT TOWARD ENEMY pointing toward the enemy or the desired sector of fire (kill zone).

BACKBLAST AREA OF M18A1 (CLAYMORE)

A Claymore's backblast can cause injury by concussion in an area 16 meters (53 ft) to the rear and sides of the mine. It can injure you by fragmentation in an area 100 meters (328 ft) to the rear and sides of the mine. You should not be within 16 meters (53 ft) of the rear of the mine. If within 100 meters (328 ft) of the rear of the mine, you should be under cover.

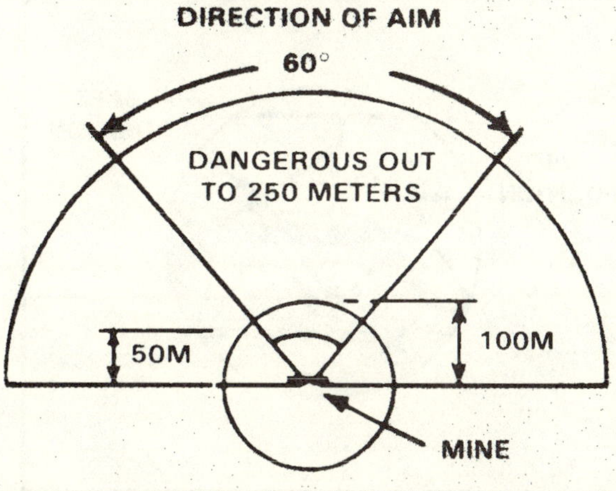

FIRING A CLAYMORE MINE

To disarm and remove a Claymore, reverse arming and emplacing procedure.

- Make sure that the firing-device safety bail is in the SAFE position.

FIRING-DEVICE SAFETY BAIL

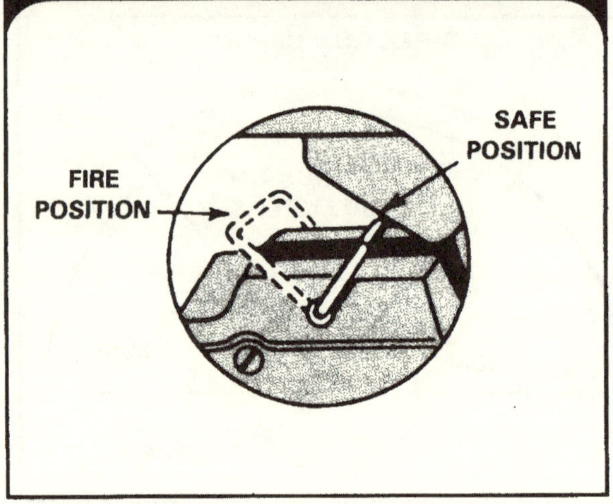

- Disconnect the firing wire from the firing device and replace the dust covers.
- Move to the mine and unscrew and remove the shipping-plug priming adapter from it. Take the firing device with you.
- Remove the blasting cap from the shipping-plug priming adapter and then screw the adapter back into the detonator well.
- Put the blasting cap inside its cardboard container, remove the firing wire from the stake, and reroll the firing wire.
- Pick up the mine and put it in the bandoleer.
- Remove the firing wire from the stake at the firing position and put it in the bandoleer.

To emplace a Claymore with a **tripwire:**

- Emplace and aim the mine to cover the desired kill zone.
- Put an anchor stake (1) about 1 meter (3.3 ft) to the rear of the mine and attach the firing wire to it, leaving about 1.5 meters (5 ft) of slack. Do not insert the blasting cap into the mine at this time.
- Unroll the firing wire to a point about 20 meters (66 ft) to either the left or right front of the mine. Put an anchor stake (2) at that point.
- Attach a clothespin (or other improvised device) to stake (2) with its closed end pointing toward the kill zone. The clothespin can be tied or nailed to the stake.
- Move across the kill zone and put in another stake (3).
- Attach the trip wire to stake (3) and unroll the tripwire to stake (2).

EMPLACING A CLAYMORE WITH A TRIPWIRE

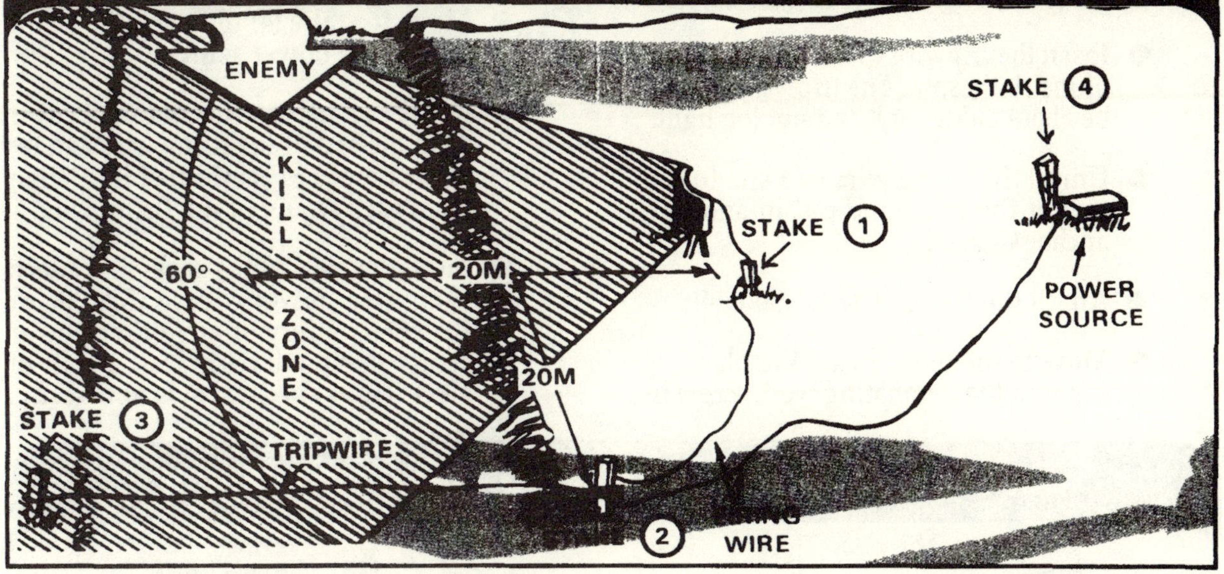

- Attach the end of the tripwire to a C-ration plastic spoon or some other nonconductor of electricity. Connect the tripwire and spoon before setting up the mine.
- Prepare the firing wire at stake (2) for connection to the clothespin by cutting one strand of the firing wire and forming two bare wire loops that can fit over the ends of the clothespin. (Do that before setting up the mine.)

FIRING WIRE PREPARED FOR CONNECTION

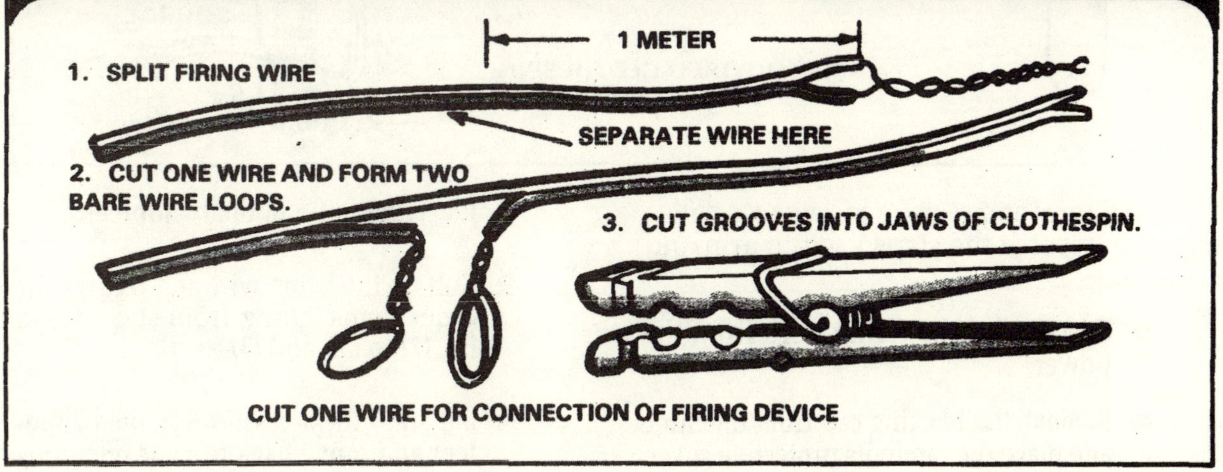

- Slide the loops over the ends of the clothespin and tighten them to fit in the grooves of the clothespin.

- Insert the tripwire spoon into the jaws of the clothespin. The tripwire should be about ankle-high and not too tight.

- Unroll the firing wire to a site to the rear of the mine and put in another anchor stake (4).

- Attach the firing wire to stake (4).

- Move to the mine, insert the blasting cap into the detonating well, screw in the shipping-plug priming adapter, and recheck the aim.

- Move to stake (4) to attach the firing wire to the power source.

- Cut the shorting plug and dust cover from the end of the firing wire and remove about 2.54 cm (1 in) of insulation from each strand of the firing wire.

- Twist the ends of the wires and attach them to a power source (BA 206 or BA 4386 battery or any other power source that produces at least 2 volts of electricity). The system is now ready.

TRIPWIRE CONNECTION TO SPOON AND CLOTHESPIN

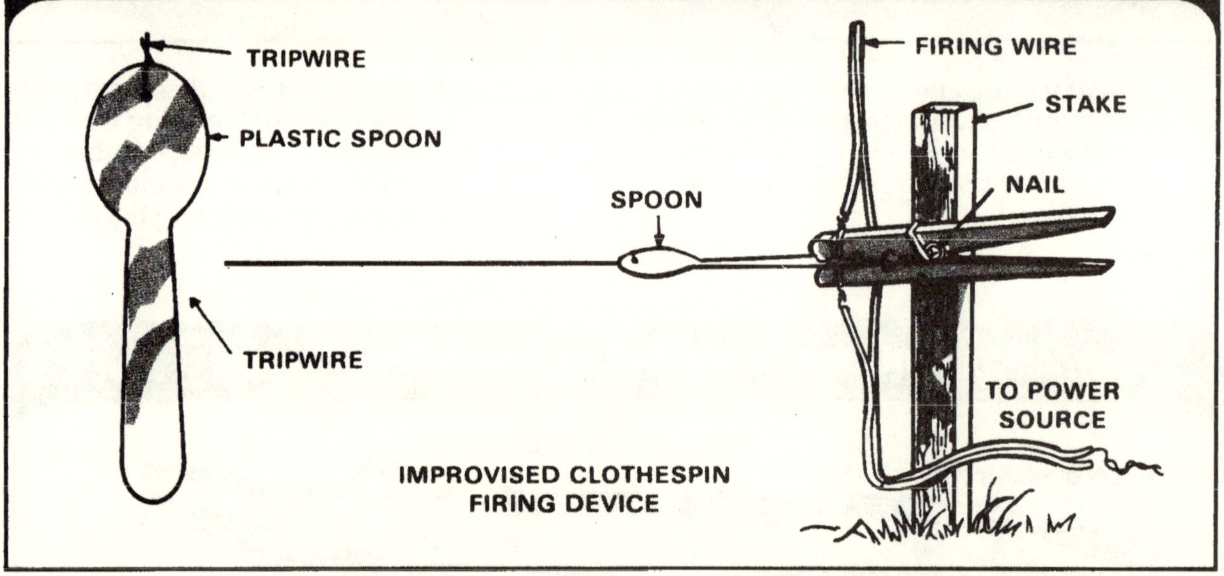

IMPROVISED CLOTHESPIN FIRING DEVICE

To disarm and remove a Claymore with a tripwire, reverse the steps used to arm and install it.

- Disconnect the firing wire from the power.

- Remove the blasting cap from the mine and place the cap in its protective cover.

- Place the mine in the bandoleer.

- Roll up the firing wire and recover the other items, going from stake (1) to (2), (2) to (3), and (3) to (4).

- Put all of the accessories in the bandoleer and move back to your position.

M26, ANTIPERSONNEL MINE

This is a small, bounding, fragmentation mine. It can be set for either pressure or tripwire activation. A pressure of 13 kg (28 lb) on top of the mine, or a pull against the tripwire will detonate it.

M26 ANTIPERSONNEL MINE

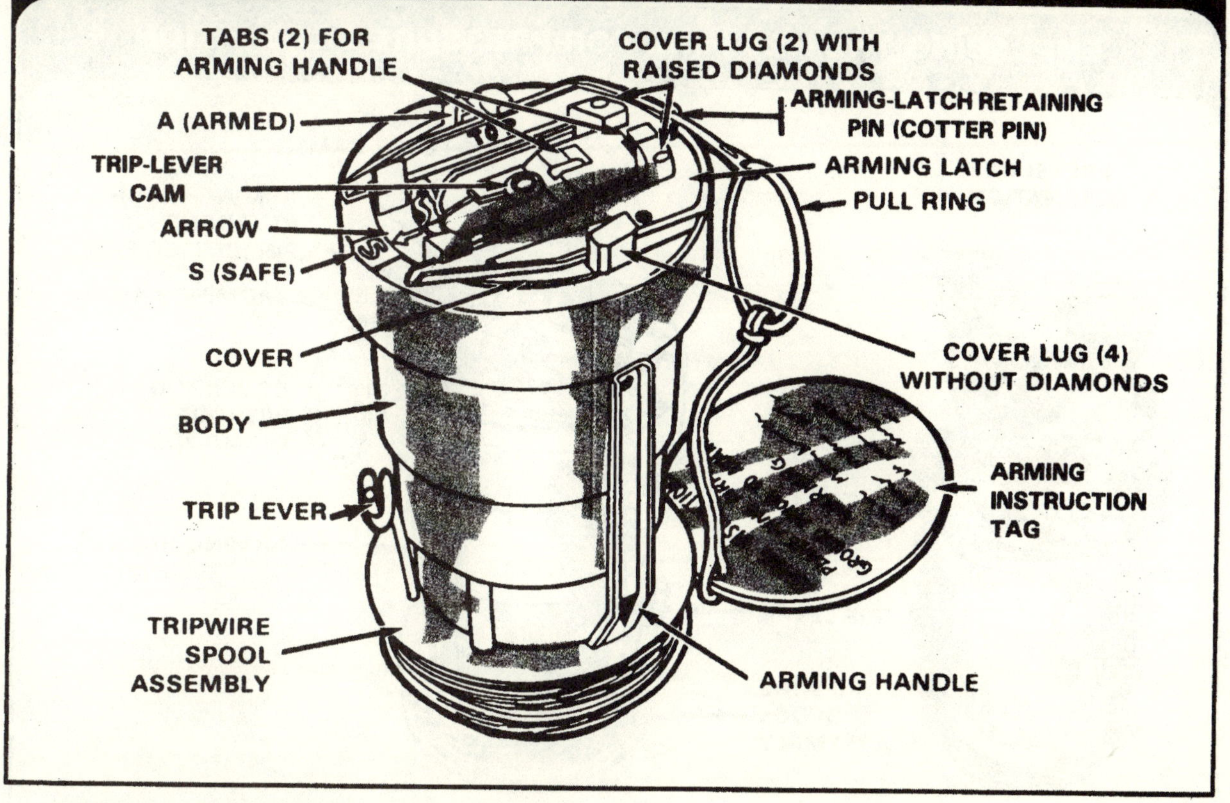

To emplace an M26 mine for **pressure** detonation:

- Dig a hole in the ground about 13 cm (5 in) deep and wide enough to accept the mine.

- Remove a 2.5 cm (1 in) layer of dirt out to about 15 cm (6 in) from around the mine to allow knuckle clearance needed when turning the arming handle.

> NOTE: Leave the tripwire spool assembly attached to the mine This helps to stabilize the mine in the hole.

- Remove the arming handle from the tripwire spool by pulling it upward from the spool.

EMPLACING THE M26 MINE FOR PRESSURE DETONATION

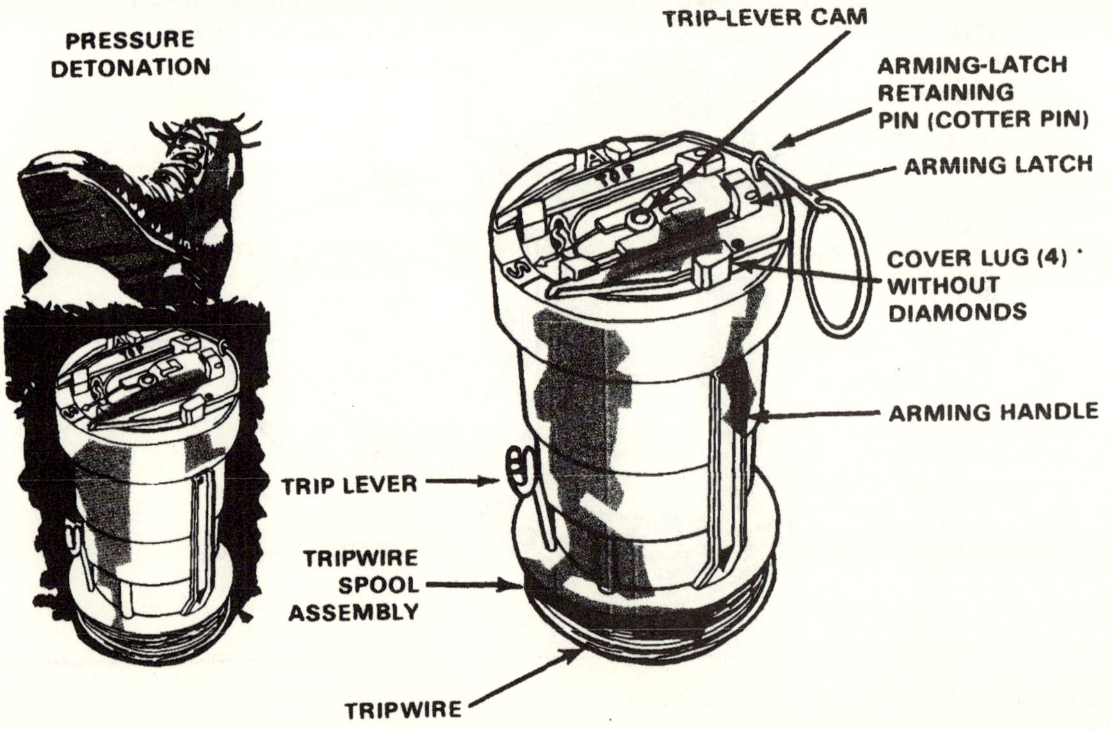

- Close the spread ends of the arming-latch retaining pin (cotter pin) to aid its removal after the mine is emplaced.

- Place the mine in the hole, with cover end up, so that the mine cover lugs extend just slightly above ground level.

- Pack the dirt around the mine, leaving the cover lugs exposed.
- Remove the arming-latch retaining pin by pulling the ring straight up.
- Attach the arming handle to the lugs on the arming latch. Hold the mine firmly with the thumb and finger of one hand to keep it from turning. Turn the cover clockwise (about ¼ turn) until it stops.
- The arrow on the mine cover should be slightly past the center of the red A (armed) position.
- Camouflage the mine.
- Remove the arming latch from the mine by pulling straight out on the arming handle. Keep the arming latch and arming handle for future use. The mine is now armed.

ARMING THE M26 MINE

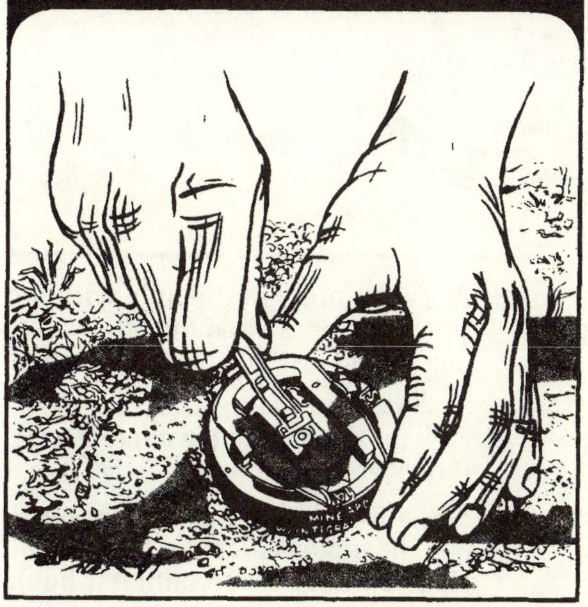

To disarm and remove an M26 mine set for pressure detonation, reverse the steps used to arm and install the mine.

- Carefully remove all camouflage from around the top of the mine.
- If there is any sign of boobytrapping or tampering, do not try to disarm and remove the mine. Instead, destroy the mine in place.
- If there are no signs of boobytrapping or tampering, replace the arming latch by sliding it under the arming handle lugs from the side opposite the arrow.
- Make sure that the middle prong of the arming latch engages the trip-lever cam.
- Remove a 2.5 cm (1 in) layer of dirt out to about 15 cm (6 in) from the edge of the mine to allow knuckle clearance.
- Attach the arming handle to the arming latch.
- Hold the mine with one hand and turn the cover counterclockwise with your other hand until it stops (about ¼ turn).
- The arrow on the cover should line up with the S (SAFE) position on the mine.
- Remove the arming handle.
- Insert the arming-latch retaining pin through the holes in the arming latch and mine body. (It may be necessary to rotate the latch back about ½ cm [¼ in] to align the holes.)
- Remove the mine from the hole.
- Clear the mine and repackage it.

To emplace an M26 mine for **tripwire** detonation:

- **Dig a hole in the ground about 13 cm (5 in) deep and wide enough for the mine.**
- **Remove a 2.5 cm (1 in) layer of dirt out to about 15 cm (6 in) from around the mine to allow knuckle clearance needed when turning the arming handle.**
- **Remove the tripwire spool assembly by pulling it away from the mine body.**
- **Remove the arming handle from the tripwire spool by pulling it upward.**
- **Unscrew and remove the trip lever from the tripwire spool.**
- **Remove one or more tripwires, as required, from the tripwire spool by pressing in on the plastic tripwire retainer(s) and lifting the tripwire(s) off the top of the spool.**

EMPLACING THE M26 MINE FOR TRIPWIRE DETONATION

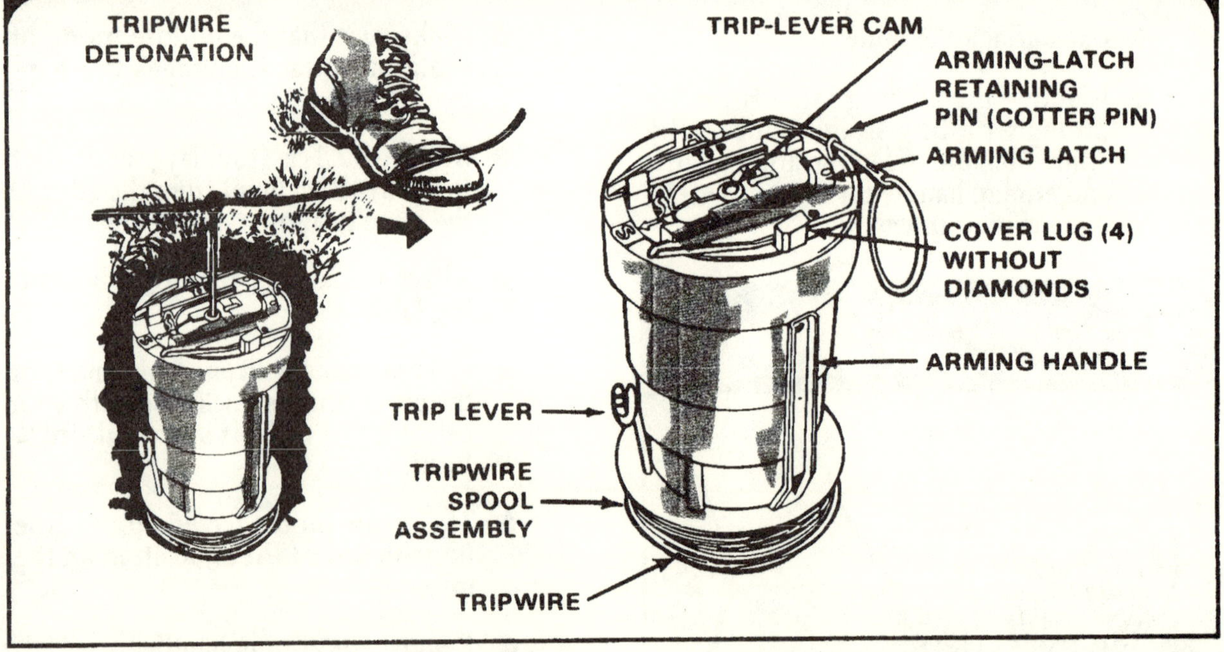

- **Retain or replace any unused tripwires on the spool.**
- **Replace the tripwire spool assembly on the mine. Leave the spool on the mine to help stabilize it.**
- **Close the spread ends of arming-latch retaining pin (cotter pin) to aid its removal after the mine is emplaced.**
- **Place the mine in the hole with the cover end up, so that the mine cover lugs extend just slightly above ground level.**
- **Pack the dirt around the mine, leaving the cover lugs exposed.**
- **Screw the trip lever about four turns into the trip-lever cam (in the top center of the mine cover) until it is tight.**

- Cut the tape holding the coils of wire.
- Attach the loop end(s) of the tripwire(s) to the loop of the trip lever.
- Leaving some slack, attach the other end of the tripwire(s) to a firm anchor stake(s).
- Remove the arming-latch retaining pin by pulling the ring straight up.
- Assemble the arming handle to the lugs on the arming latch. Hold the mine body firmly with your thumb and finger of one hand to keep the mine from turning. Rotate the cover clockwise (from S to A) until it stops (about ¼ turn).
- Camouflage the mine.
- Remove the arming latch from the mine by pulling straight out on the arming handle. Keep the arming latch and arming handle for future use. The mine is now armed.

To disarm and remove an M26 mine emplaced for tripwire detonation, reverse the steps used to arm and install the tripwire detonation.

- Carefully remove all camouflage from around the mine.
- If there is any evidence of boobytrapping or tampering, do not try to disarm and remove the mine. Take care not to move the trip lever or press on the mine cover.
- Replace the arming latch. With the two raised arming handle lugs facing upward, slide the arming latch under the six lugs of the mine cover from the side opposite the arrow. Make sure that the middle prong of the arming latch engages the trip-lever cam.
- Remove a layer of dirt about 2.5 cm (1 in) deep for a distance of about 15 cm (6 in) from the edge of the mine to allow knuckle clearance for turning and removing the arming latch.
- Attach the arming handle to the lugs on the arming latch.
- Hold the mine with one hand and turn the mine cover counterclockwise with your other hand until it stops (about ¼ turn). The arrow on the cover should point to the S (SAFE) position on the mine body.
- Remove the arming handle from the mine and keep it for future use.
- Insert the arming-latch retaining pin through the holes in the arming latch and mine body. It may be necessary to turn the latch back (up to ½ cm [¼ in]) to align the holes of the latch and body.
- Remove the mine from the ground. Clean the mine and repackage it.

ANTITANK

M15, ANTITANK MINE

This antitank mine has a cylindrical steel body. It is pressure detonated. A force of 159 to 340 kg (350 to 750 lb) on the pressure plate will detonate the mine.

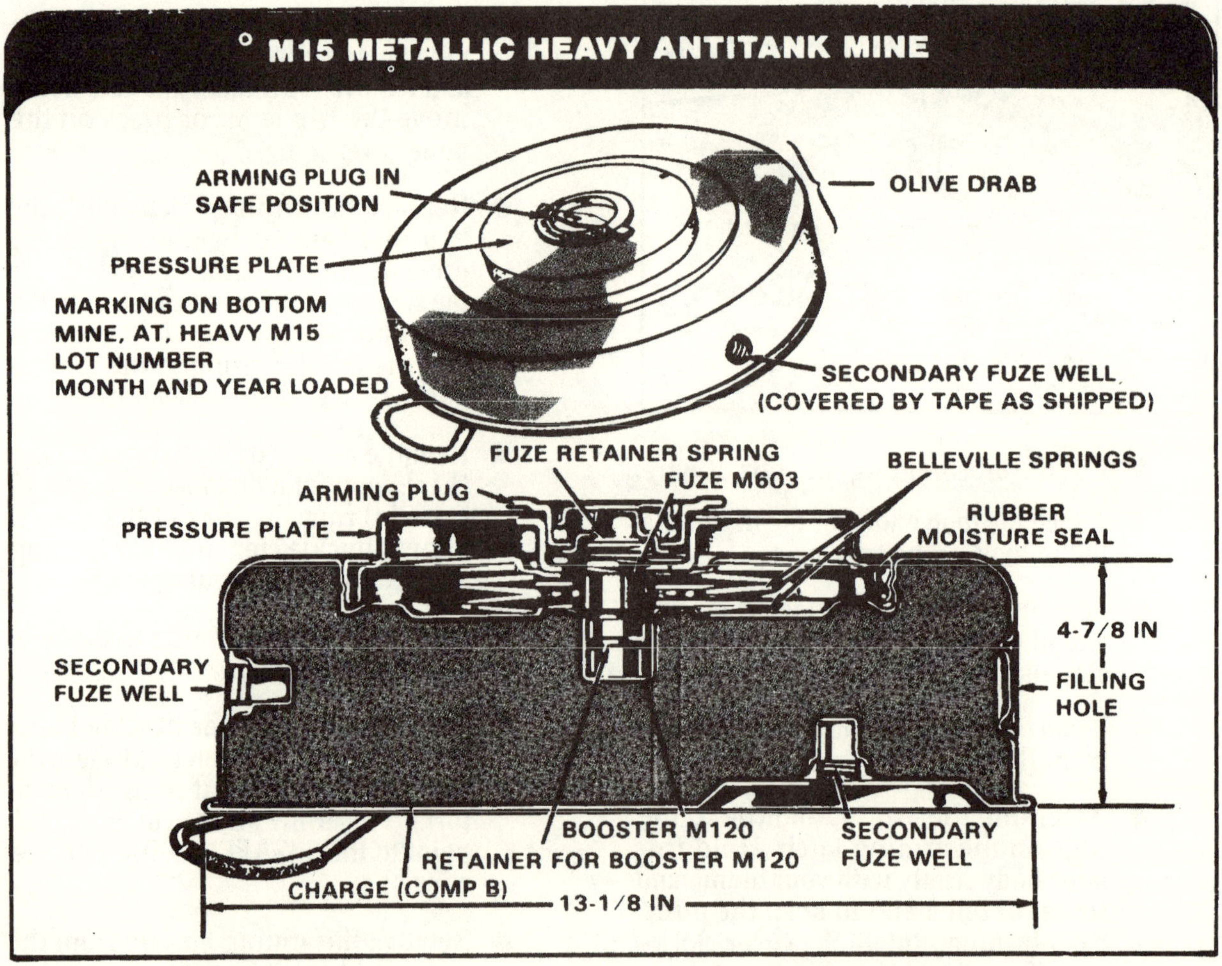

To emplace an M15 mine:

- Remove the mine from its packing box.
- Using the M20 wrench, unscrew the arming plug by turning it counter-clockwise. Take it out of the mine.
- Inspect the fuze well for foreign matter. Remove any found.

ARMING WRENCH M20

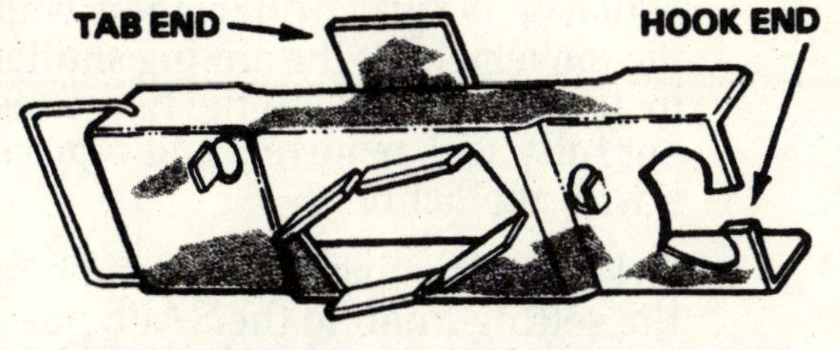

REMOVING ARMING PLUG FROM MINE

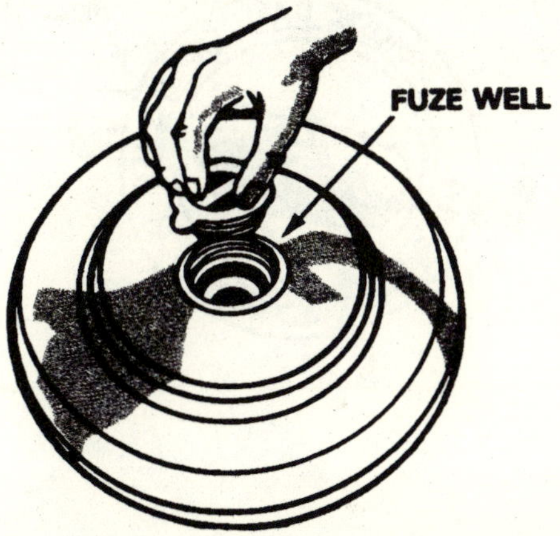

- Make sure that the booster retainer is seated in the fuze well. If it is missing, replace the mine.
- Put the mine down and pick up the metal fuze container.
- Open the container with the key attached to its bottom.
- Remove the fuze from the container.
- Make sure that the green end of the detonator shows in the bottom of the fuze and that the safety clip is in place between the pressure plate and the body of the fuze.

CONTAINER AND MINE FUZE M603

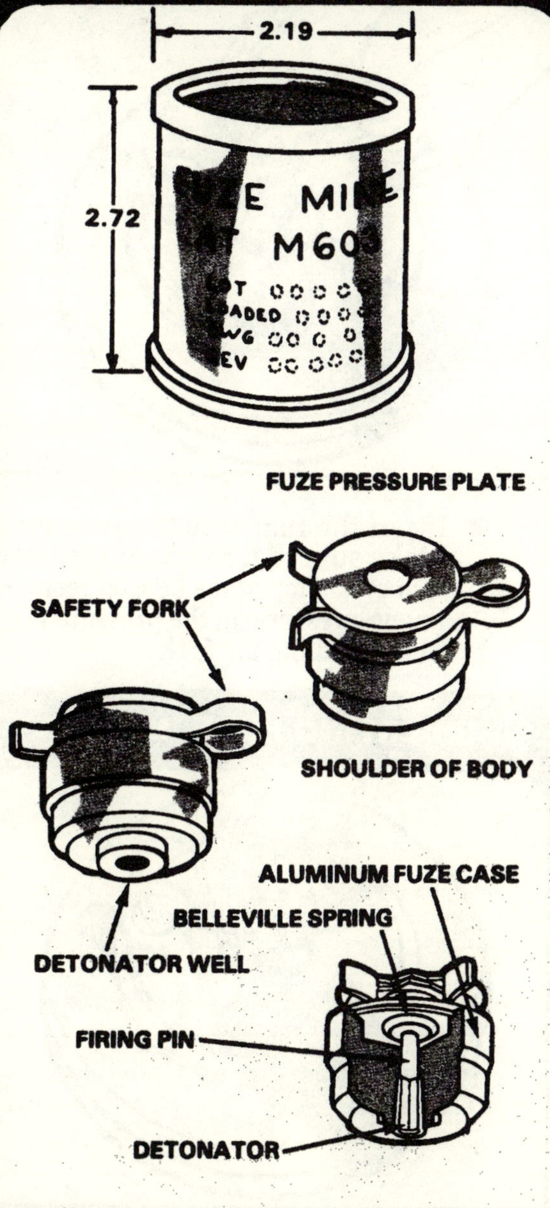

- Remove the safety fork from between the pressure plate and the body of the fuze. Keep the safety fork for future use.

REMOVING SAFETY FORK FROM FUZE

- Insert the fuze into the fuze well. Make sure that the fuze is seated securely on top of the booster retainer. Put no-pressure on the pressure plate when handling the fuze.

INSERTING FUZE IN MINE

- Check the clearance of the pressure plate of the fuze in the fuze well by using the tab end of the M20 wrench. If the pressure plate is too high, the button on the plate will interfere with the movement of the arming shutter in arming the mine. If the fuze does not fully seat, remove it and replace it with another fuze.

- Pick up the arming plug M4 and turn the setting knob to the SAFE position — if it is not already on SAFE.

ARMING PLUG AND WELL

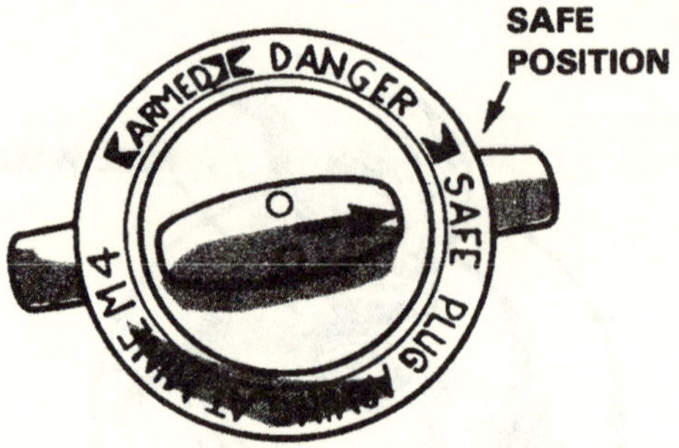

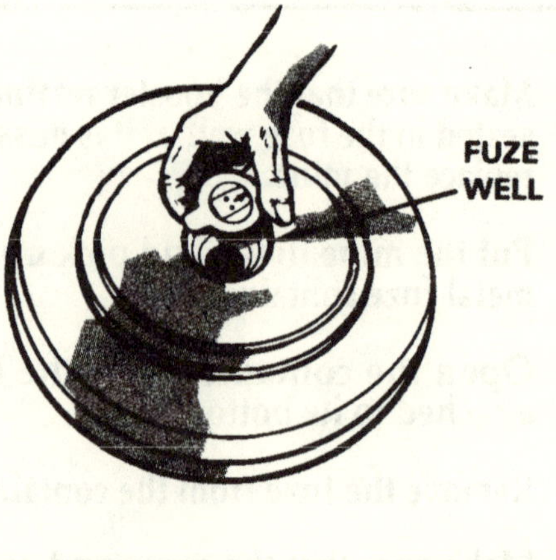

RECOMMENDED BURIAL FOR PRESSURE FUZED MINES

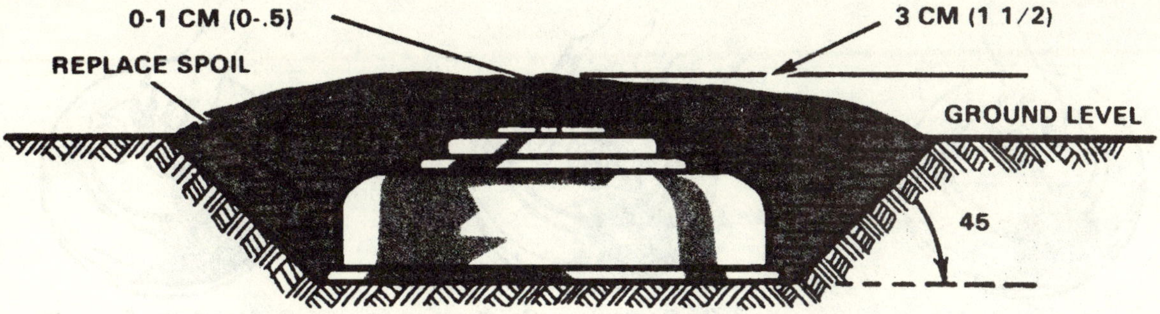

- Dig a hole about 38 cm (15 in) in diameter and 15 cm (16 in) deep, with walls sloping 45 degrees.
- Check the bottom of the hole to make sure that the ground is solid so that the mine will not sink into the ground. If it is not solid, insert a wooden board or other support to give the mine a firm foundation.
- Lay the mine in the hole so that the top surface of the pressure plate is about 3 cm (11/2 in) below ground level.
- Fill in the dirt around the mine and pat it down.
- Using the M20 wrench, arm the mine by turning the setting knob from SAFE through DANGER to ARMED.
- Camouflage the mine.

ARMING THE M15 MINE

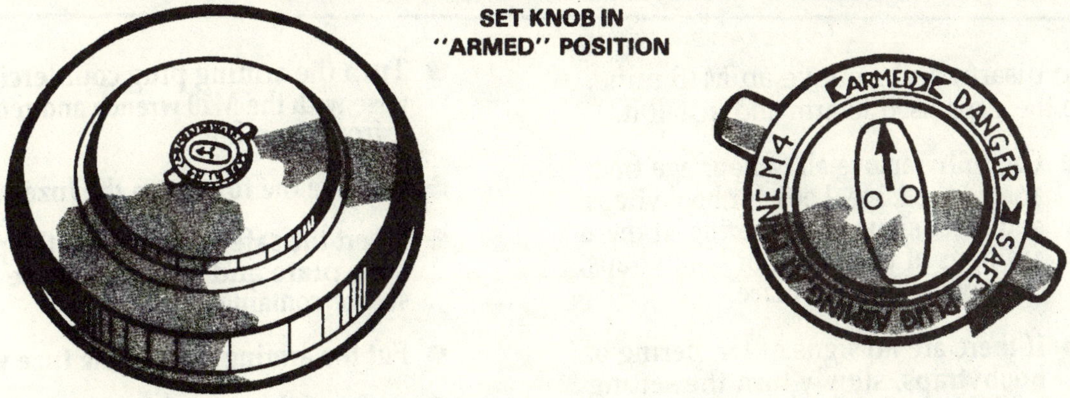

STEPS IN ARMING THE M15 MINE WITH FUZE AT M603

1 Removing Arming Plug From Mine.

2 Removing Safety Fork From Fuze.

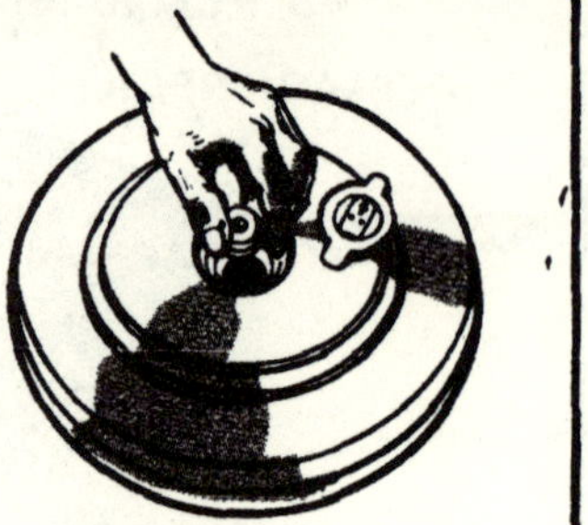

3 Inserting Fuze In Mine.

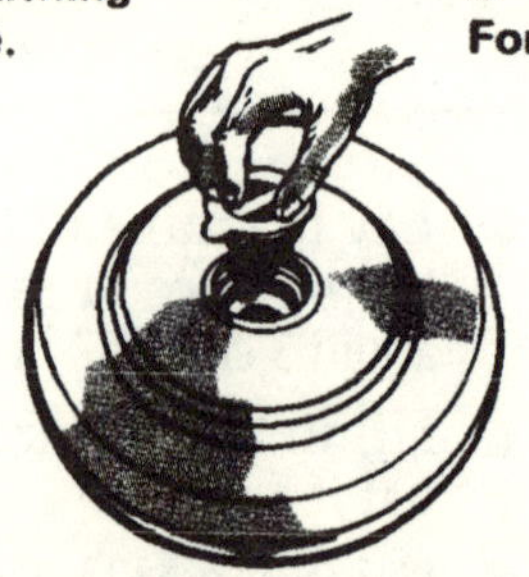

4 Inserting Arming Plug With Indicator At "SAFE" Position Into Mine Before Laying.

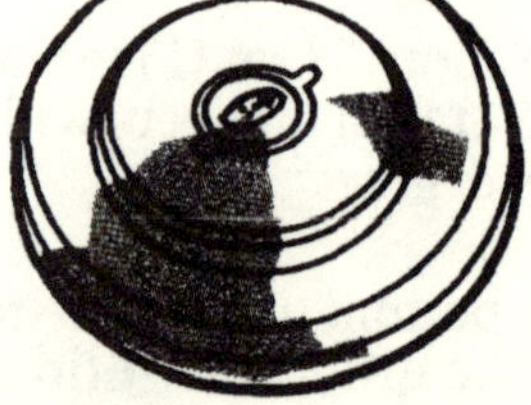

5 Indicator Turned To "ARMED" Position After Laying Mine.

WARNING

Ice in the fuze well during fuzing operations can cause a serious accident. During freezing weather, make sure none is present.

To disarm and remove an M15 mine, reverse the steps used to arm and install it.

- Carefully remove all camouflage from around the mine. Look for boobytraps and other signs of tampering. If there are signs of tampering or boobytraps, destroy the mine in place.
- If there are no signs of tampering or boobytraps, slowly turn the setting knob from ARMED through DANGER to SAFE. Use the M20 wrench.
- Turn the arming plug counterclockwise with the M20 wrench and remove it from the mine.
- Remove the fuze from the fuze well.
- Insert the safety fork under the pressure plate and place the fuze in a secure container.
- Put the arming plug in the fuze well.
- Remove the mine from the hole and put it in the packing box.

M21, ANTITANK MINE

This antitank mine has a cylindrical steel body. It is pressure detonated. A pressure of 1.7 kg (3.75 lb) against the tilt rod (causing the rod to tilt 20 degrees or more) will detonate the mine. When not using the tilt rod, a pressure of 131.5 kg (290 lb) on the pressure ring will detonate the mine.

ANTITANK MINE, M21 AND COMPONENTS

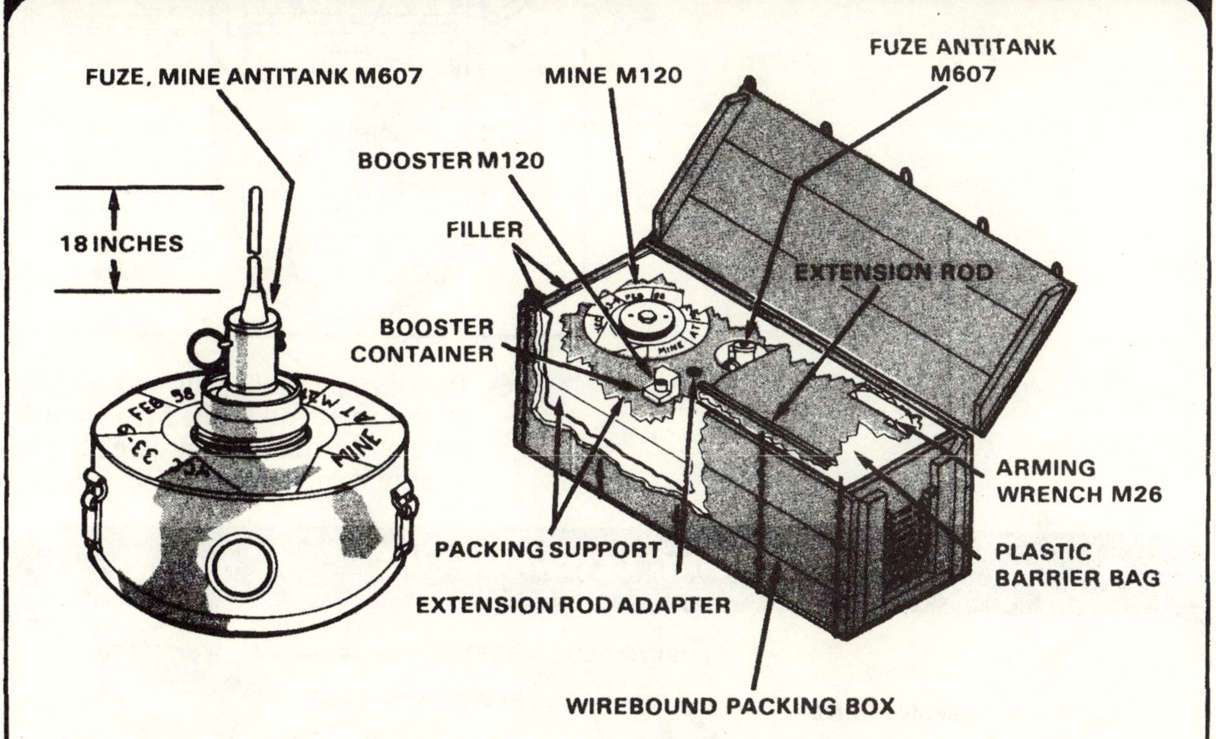

To emplace an **M21 mine:**

- **Remove the mine and its components from the packing box.**
- **Inspect the mine and components for serviceability. Check for cracks, dents, or other signs of damage. If a damaged item is found, replace it.**

- Make sure that the cotter pins of the fuze pull-ring assembly and the fuze closure assembly are in place and secure.

PIN AND FUZE CLOSURE ASSEMBLY

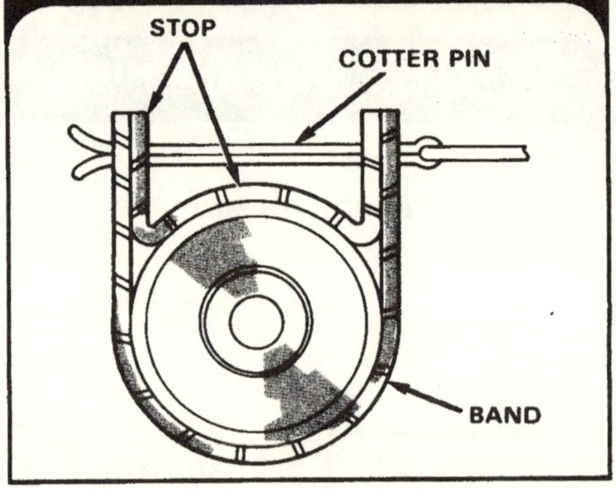

- Turn the mine bottom up and, with the screwdriver end of the M26 wrench, remove the closing plug assembly by turning it counterclockwise.

REMOVING CLOSING PLUG

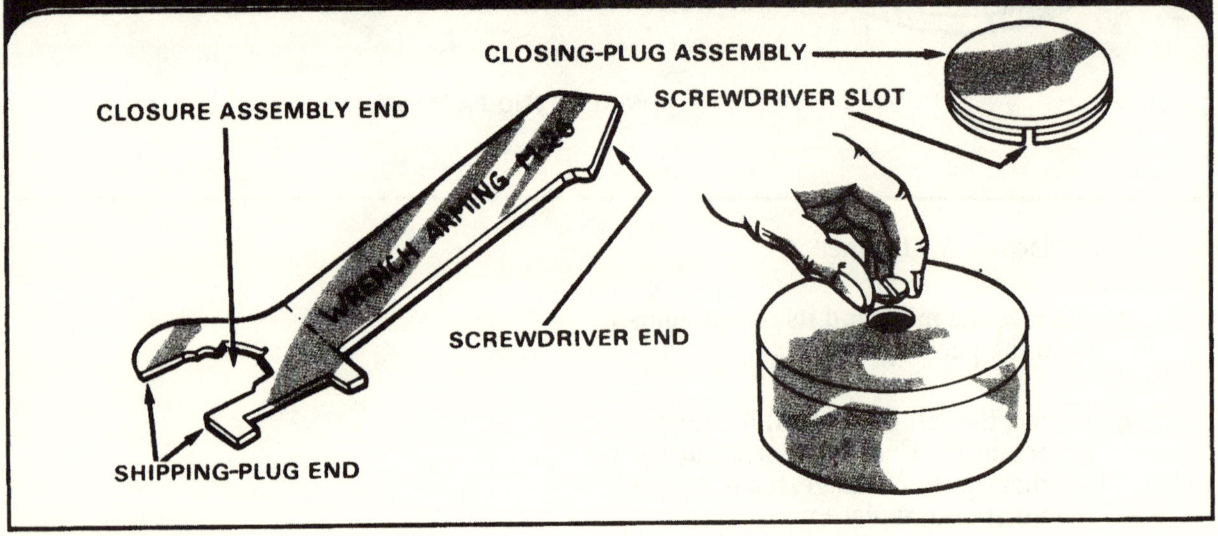

- Inspect the booster cavity for foreign matter. Remove any found.
- Insert the M120 booster (with the washer side toward the fuze) into the booster cavity.
- With the M26 wrench, replace the closing-plug assembly by turning it clockwise until tight. The gasket of the closing-plug assembly should be against the booster.
- Turn the mine bottom down.
- With the M26 wrench, remove the shipping-plug assembly from the fuze hole of the mine.

MINE ANTITANK HEAVY M21 (INTERNAL) CUTAWAY VIEW

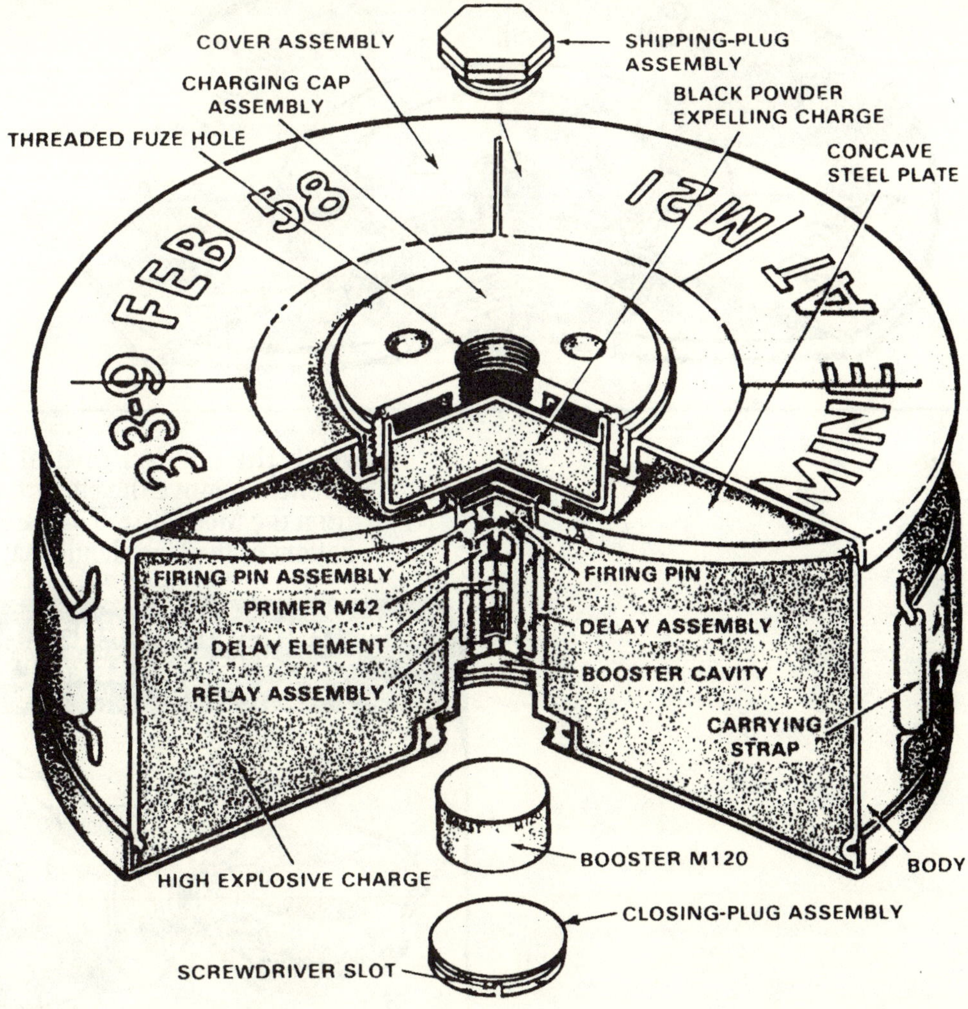

- Inspect the fuze hole. If foreign matter is present, remove it from the hole.

REMOVING THE SHIPPING PLUG

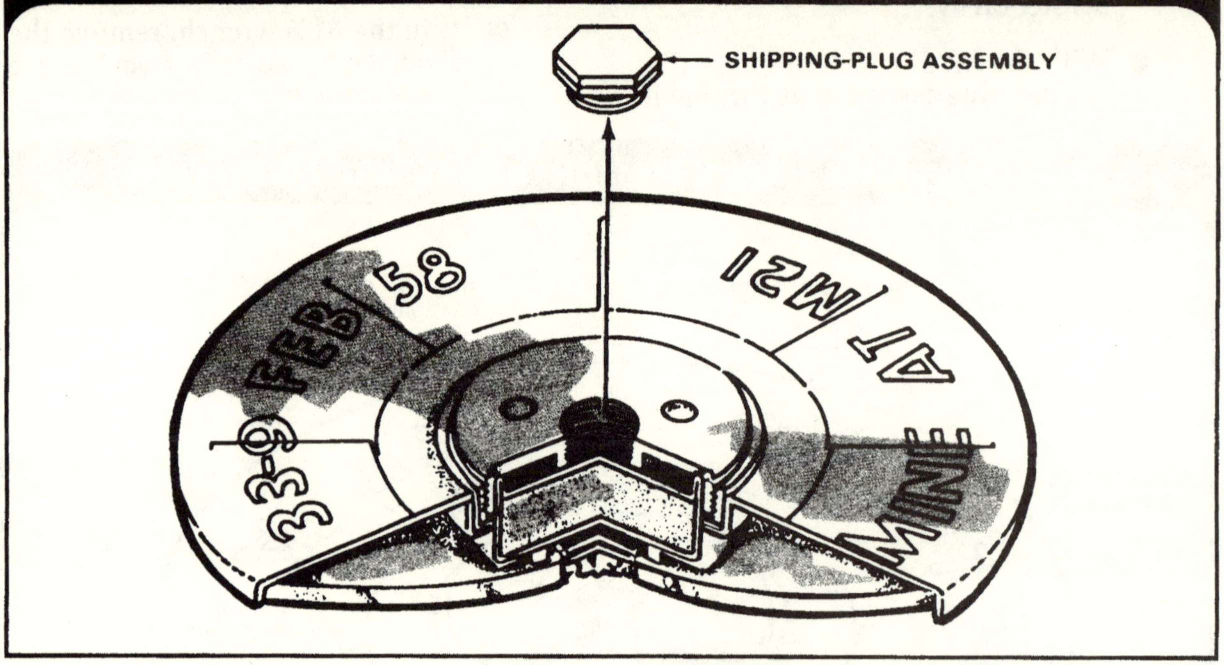

- With the closure end of the M26 wrench, remove the closure assembly from the M607 fuze. The gasket on the bottom of the fuze should stay in place.

DISCONNECTING FUZE FROM CLOSURE ASSEMBLY

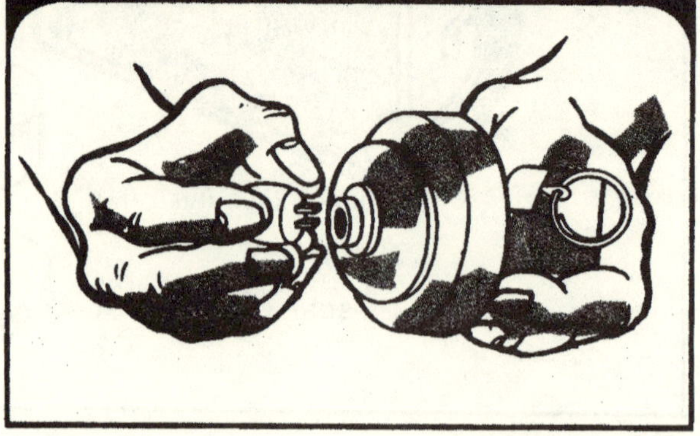

- Screw the fuze hand-tight into the threaded fuze hole of the mine charge cap. Set the mine down.

FUZE IN HOLE

- Dig a hole in the ground 30 cm (12 in) in diameter and 15 cm (6 in) deep.
- Check the bottom of the hole to make sure that the ground is solid and has a firm, flat foundation for the mine to rest on. If the ground is soft, the mine may tilt and lose effectiveness.
- In soft ground, place a board or other flat object under the mine as a firm foundation.
- Place the mine in the hole.
- Press the ground firmly against the sides of the mine, leaving the fuze uncovered.
- Screw the extension rod into the threaded pressure ring of the fuze.
- Make sure that the extension rod is vertical.

If the mine is being set for pressure detonation with the pressure rings, do not use the extension rod. **Instead:**

- Remove the pull ring assembly band and stop on the fuze. This arms the mine.
- Keep the above items for future use, if needed, to disarm the fuze.
- Camouflage the mine.

To disarm and remove an M21 mine, reverse the steps used to arm and install it.

- Check the area for boobytraps or any signs of tampering. If there are boobytraps or signs of tampering, destroy the mine in place.
- If there are no boobytraps or signs of tampering, remove the camouflage material from around the mine.
- Reassemble the band, stop, and pull-ring assembly on the fuze so that the pressure ring is immobilized. When the cotter pin is in place, spread the ends so that it is not easily removable.
- Remove the extension rod and the extension-rod adapter, if present. Be careful not to damage them.
- Remove the dirt from around the mine and remove the mine from the hole.
- Remove the fuze from the mine and install the closure assembly on the fuze.
- Install the shipping-plug assembly into the fuze hole of the mine.
- Turn the mine bottom up and remove the closing-plug assembly.
- Remove the booster, then reinstall the closing-plug assembly with the gasket toward the booster cavity.
- Put the mine, fuze, and components in their original container.

M24, OFF-ROUTE ANTITANK MINE

This is a remotely detonated mine system. It is activated by vehicles running over a linear switch (called a discriminator) which causes a 3.5-inch HEAT (HIGH EXPLOSIVE ANTI-TANK) rocket to be launched from an "off-route" launch position. The launcher should be between 3 and 30 meters (10 to 100 feet) from the edge of the path.

EMPLACED MINE, ANTITANK, HE, M24

M24 MINE ACCESSORIES POUCH AND COMPONENTS

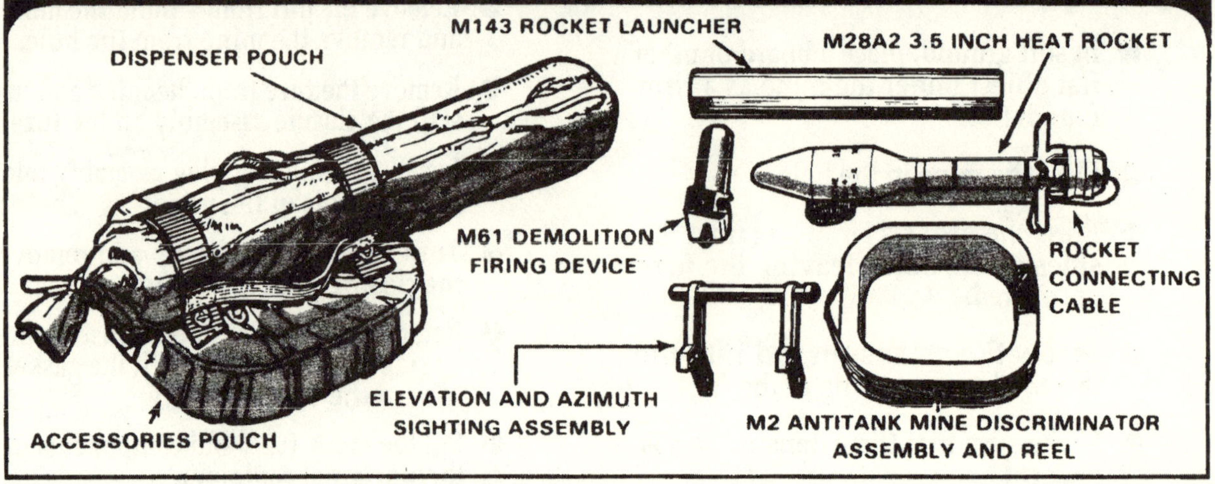

APPENDIX B

Demolitions

GENERAL

There will be times when you have to use demolitions for:

- **Breaching minefield.**
- **Breaching wire obstacles.**
- **Clearing landing zones.**
- **Blowing holes in walls of buildings.**
- **Blowing down trees to create obstacles.**

HOW TO PREPARE FIRING SYSTEMS

Information on the preparation and placement of demolition charges is in FM 5-25 and in GTA 5-10-27. This appendix covers the preparation of firing systems that are basic to all demolition work. There are two types of firing systems – NONELECTRIC SYSTEM and ELECTRIC SYSTEM.

CONTENTS

GENERAL B-1

HOW TO PREPARE FIRING SYSTEMS .B-1

NONELECTRIC SYSTEM

To prepare a nonelectric firing system, take these steps:

- *STEP 1.* Clear the cap well of a block of TNT or push a hole about the size of a blasting cap (3 cm [1% in] deep and .65 cm [¼ in] in diameter) in a block of C4 explosive.

CLEAR CAP WELL IN A BLOCK OF TNT

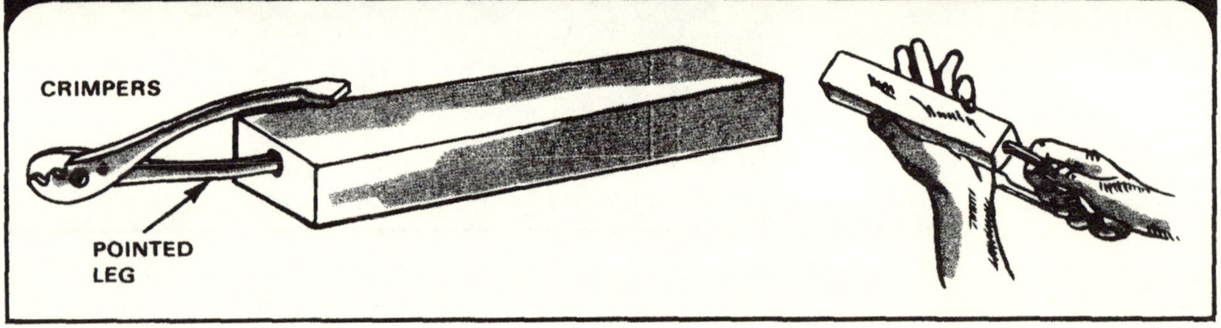

- *STEP 2.* To help prevent a misfire, cut and discard 15-cm (6-in) length of fuse from the free end of the time blasting fuse. That part of the fuse may have absorbed some moisture from the air through the exposed powder in the end of the fuse.

HELP PREVENT A MISFIRE

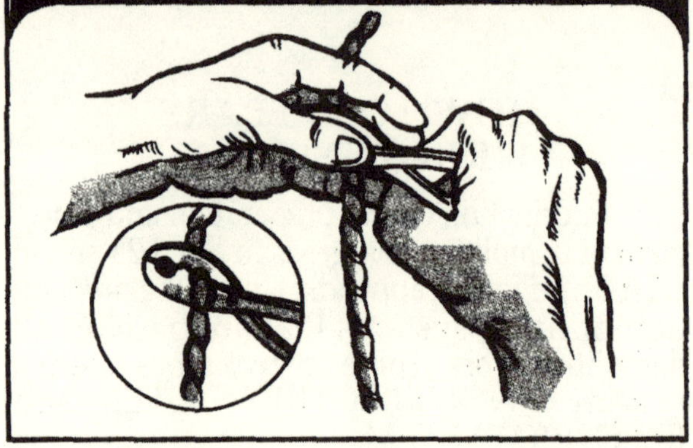

- *STEP 3.* Determine what length of fuse is needed. To do this, first compute the burning time of a 91.4-cm (3-ft) section of fuse. Divide this burning time by 3 to find the burning time of 30.5 cm (1 ft) of fuse. Next, determine the time it takes to reach a safe distance from the explosion. Now divide the time required to reach that distance by the burn time of 30.5 cm (1 ft) of fuse. This will give the number of centimeters (ft) of fuse needed.
- *STEP 4.* Inspect the nonelectric blasting cap to make sure it is clear of foreign matter.
- *STEP 5.* Gently slip the blasting cap over the fuse so that the flash charge in the cap is in contact with the end of the time fuse. DO NOT FORCE THE FUSE INTO THE CAP.
- *STEP 6.* After seating the cap, crimp it 3.2 mm (1/8 in) from the open end of the cap. Hold it out and away from your body when crimping.

CRIMPING THE CAP

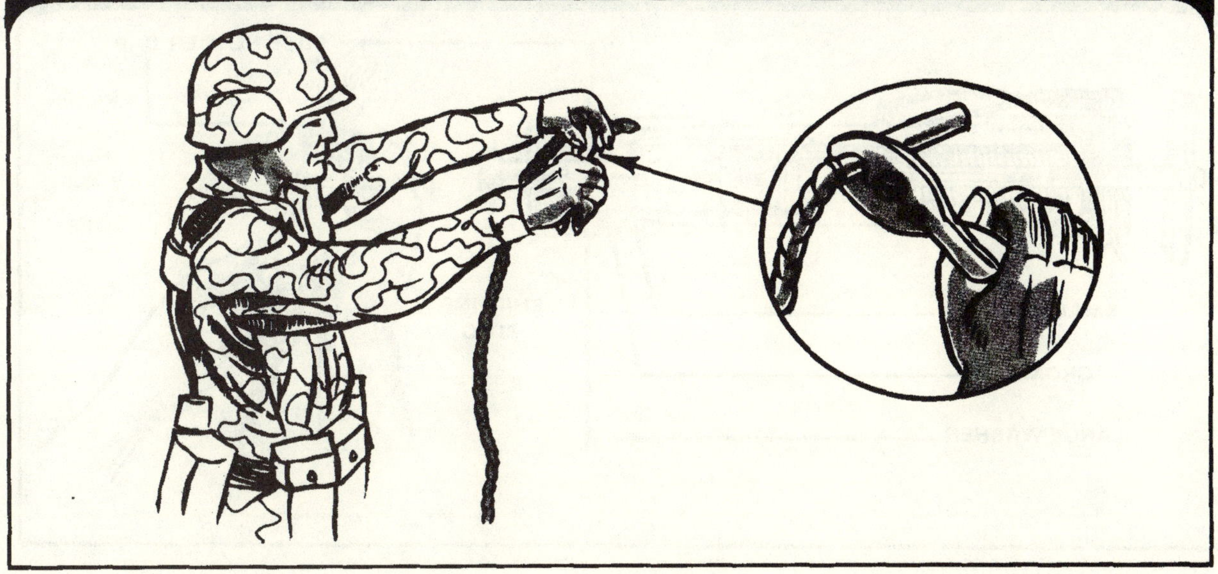

- *STEP 7.* When using TNT, insert the blasting cap into the cap well. When using C4, place the cap into the hole made in the C4 and mold the C4 around the cap. DO NOT FORCE THE CAP INTO-THE HOLE.

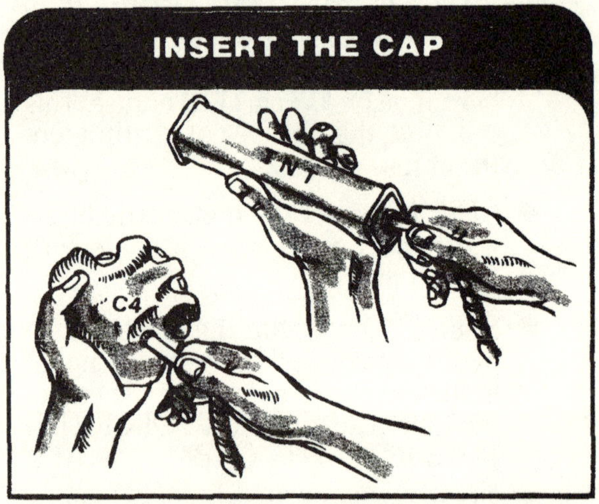

- *STEP 8.* Insert the free end of the fuse into an M60 fuse igniter and secure it in place by screwing on the fuse holder cap.

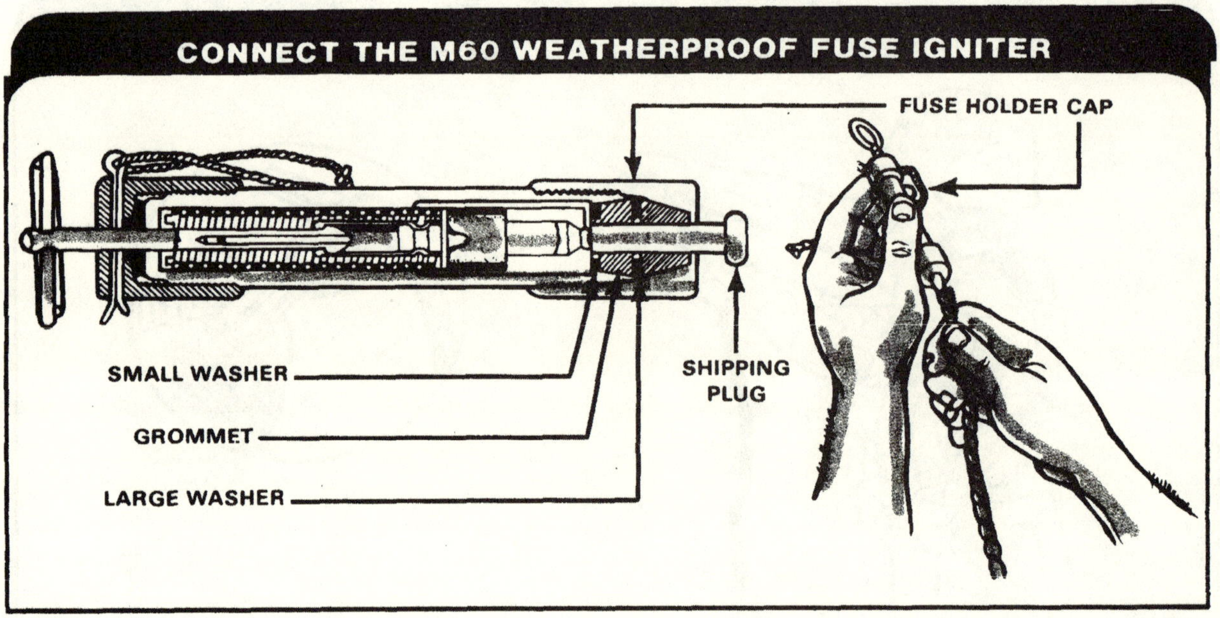

- *STEP 9.* To fire the fuse igniter, remove the safety pin, hold the barrel in one hand. Take up the slack, before making the final strong pull. If the fuse igniter misfires, reset it by pushing the plunger all the way in. Then try to fire it as before. If it still misfires, replace it.

- *STEP 10.* If a fuse igniter is not available, split the end of the fuse and place the head of an unlighted match in the split. Make sure the match head is touching the powder train.

- *STEP 11.* Then light the inserted match head with a burning match or strike the inserted match head on a matchbox.

If the fuse burns but the explosive charge does not go off, there is a misfire. Wait 30 minutes before trying to clear it. If the misfire charge was not tamped (nothing packed around it), lay another charge of at least one block of C4 or TNT beside it. If it was tamped, place at least two blocks of C4 or TNT beside it. Do not move the misfire charge. The detonation of the new charge should detonate the misfire charge.

FIRE THE FUSE IGNITER

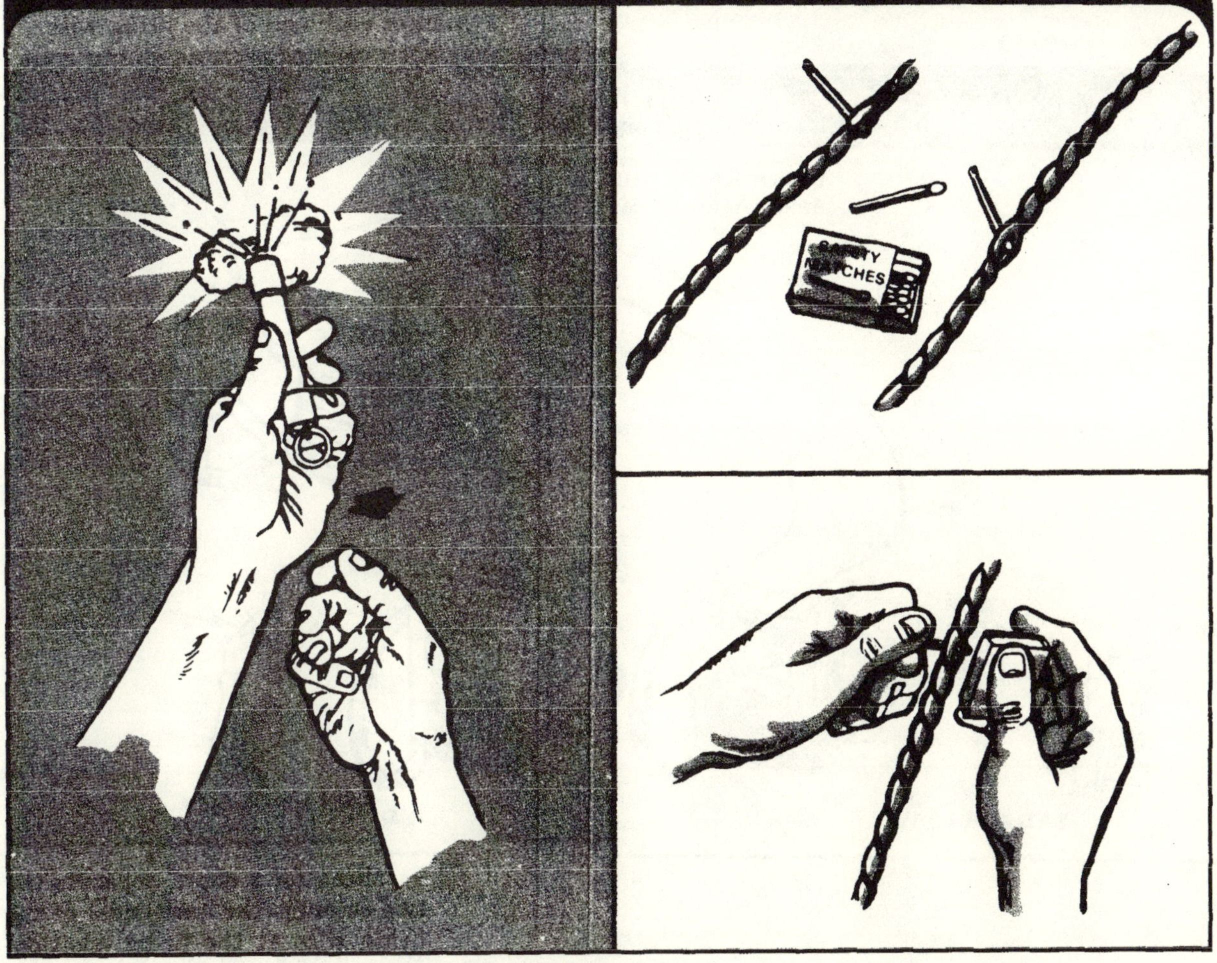

ELECTRIC SYSTEM

To prepare an electric firing system, **take these steps:**

- *STEP 1.* After finding a safe firing position and a place for the charge, lay out the firing wire from the charge position to the firing position. Before leaving the charge position, anchor the firing wire to something. Always keep the firing device with you. **Do not leave it at the firing position.**

- *STEP 2.* Check the firing wire with the galvanometers or circuit taster to make sure it does not have a short circuit or a break. This is best done with one man at each end of the firing wire.

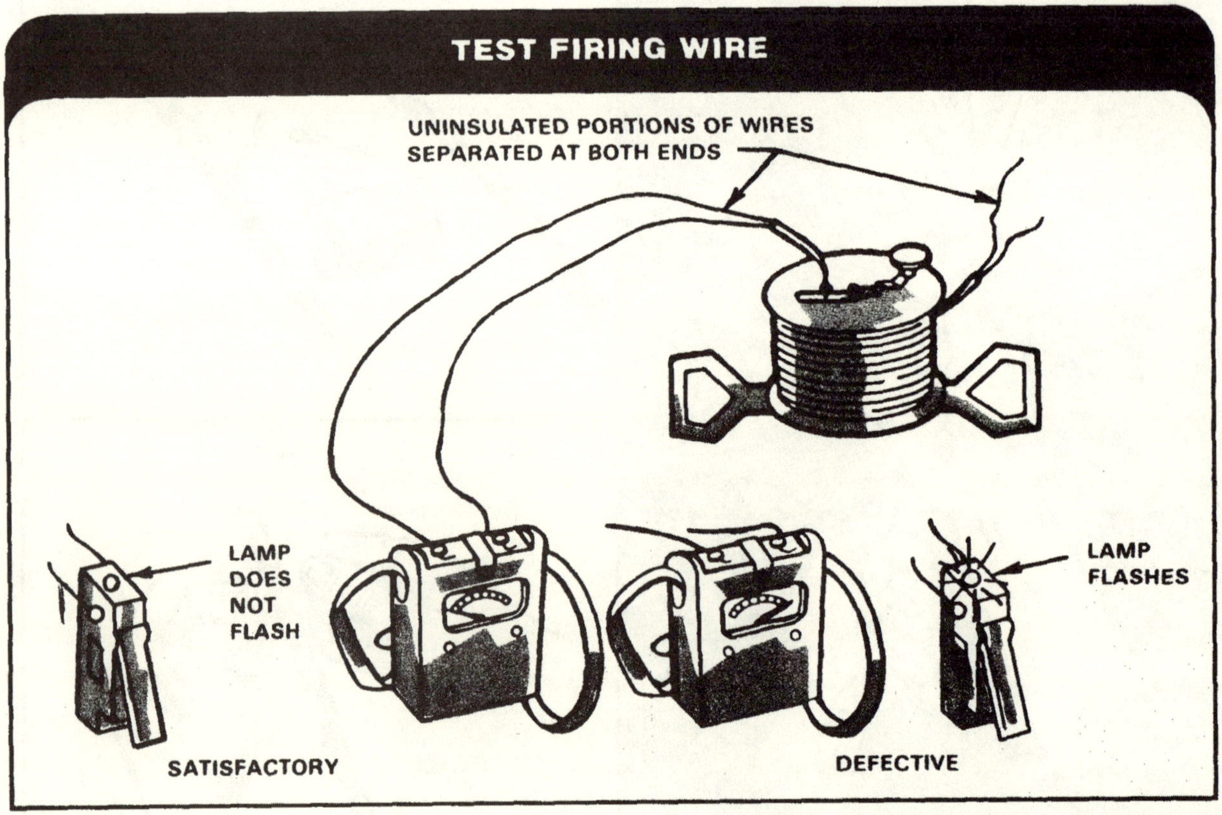

☐ To check for a short, separate the two strands (the bare ends) of the firing wire at the firing position.

Have the other soldier do the same thing with the other end of the wire at the charge position. At the firing position, touch the bare ends of the two strands to the galvanometer/circuit tester posts. The needle on the galvanometers should not move, nor should the light on the circuit tester come on. If the needle does not move or if the light does not come on, the wire has a break — **replace it.**

- ☐ If the wire has no **short** when tested, test it for a **break.** Have the soldier at the charge position twist the bare ends of the strands together. Then touch the two strands at the firing position to the galvanometers/circuit tester posts. That should cause a wide deflection of the galvanometer needle or cause the circuit tester light to come on. If the galvanometers needle does not move or if the light does not come on, the wire has a break — **replace it.**

TEST FIRING WIRE (CONTINUED)

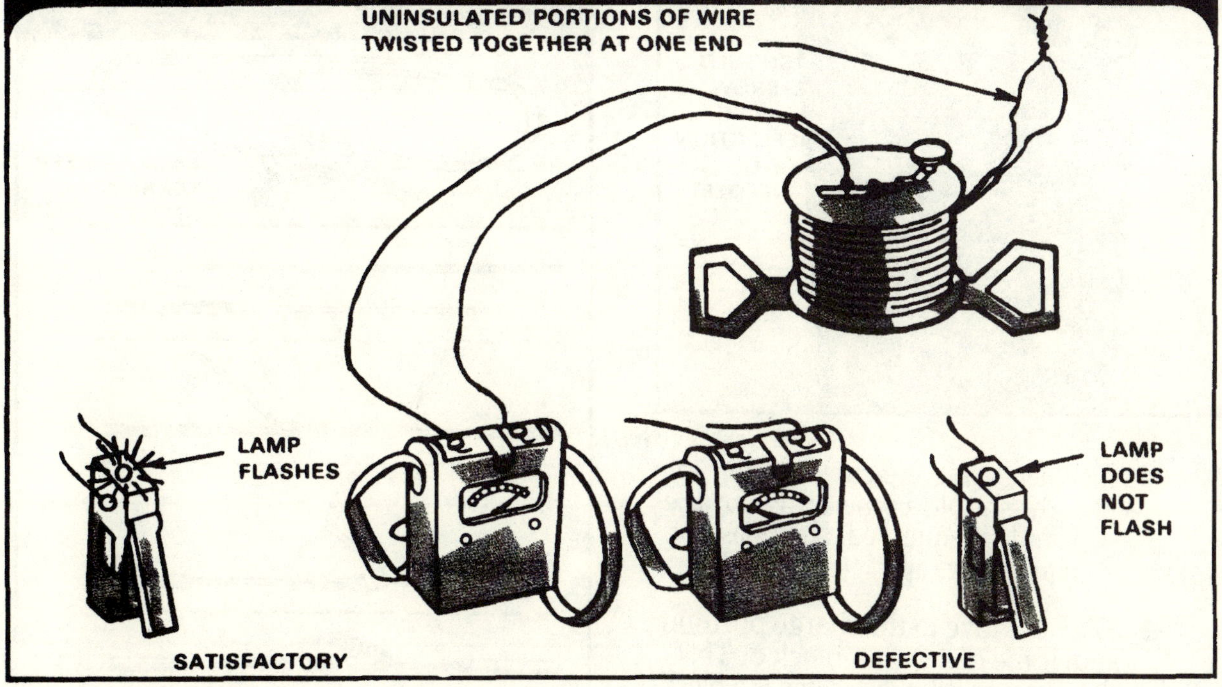

- STEP 3. At the firing position, check the blasting cap with a galvanometers or circuit tester to make sure it does not have a short. Remove the short circuit shunt and touch one cap lead wire using the galvanometers, the needle should make a wide deflection. If it does, the cap is good.

  ☐ If the needle fails to move or only makes a slight deflection, **replace the cap.**

TEST GALVANOMETER

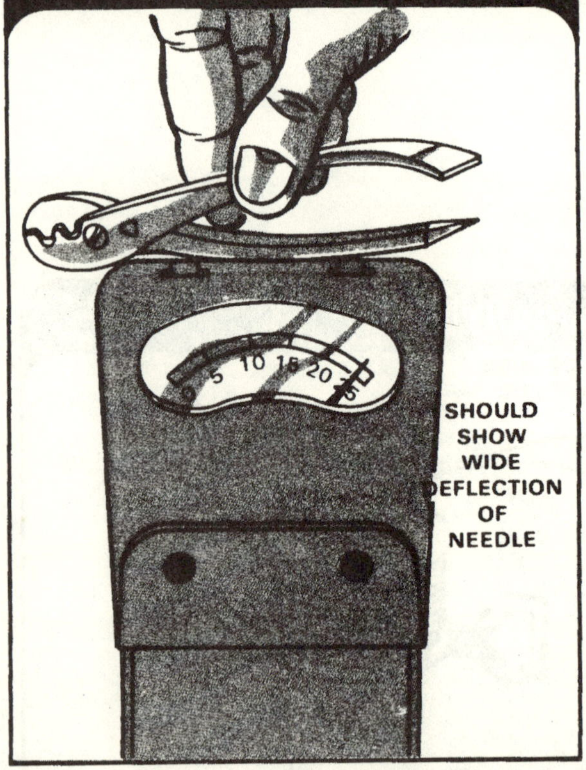

  ☐ When using the circuit tester, the light should come on when the handle is squeezed. If it does not, **replace the cap.**

- *STEP 4.* Move to the charge position and, if the charge is a block of TNT, clear its cap well if the charge is a block of C4 plastic explosive, push a hole in it about the size of a blasting cap.

TEST CIRCUIT TESTER

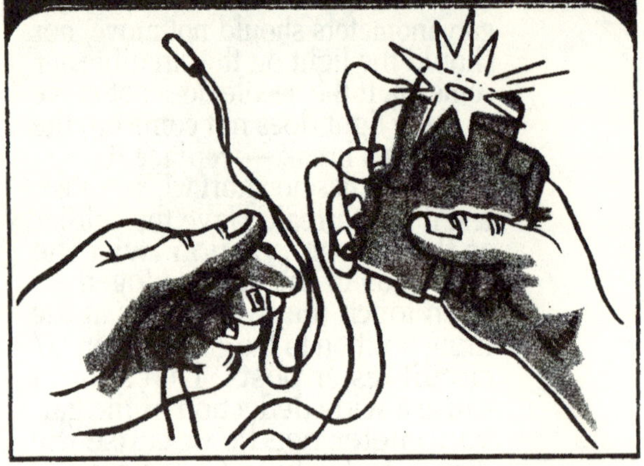

- *STEP 5.* Position the charge. Then splice the lead wires of the cap to the firing wire (pigtail knot).

PIGTAIL KNOT

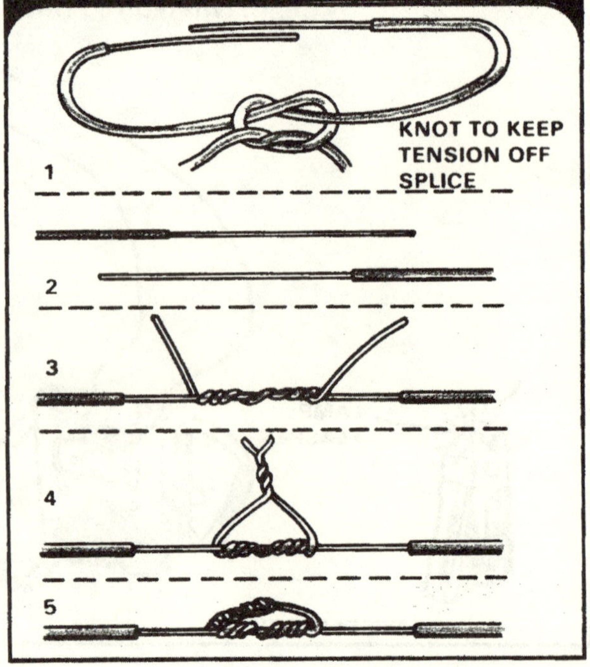

- *STEP 6.* Insert the cap into the cap well of the TNT and secure it with the priming adapter, or insert the cap into the hole made in the C4 and mold the explosive around the cap.

PRIME DEMOLITION BLOCK WITH PRIMING ADAPTER

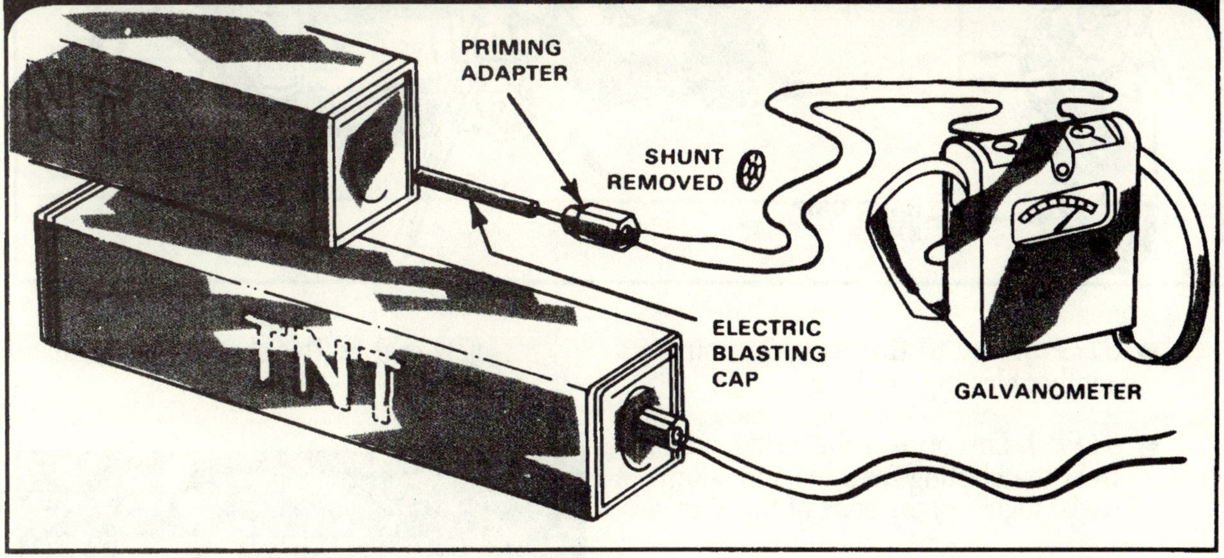

- *STEP 7.* Move back to the firing position and check the wire circuit with the galvanometers or circuit tester (same technique as described earlier).

If the circuit checked out and the blasting machine does not set off the charge, there is a misfire.

If an untamped charge misfires, investigate at once. If the charge is tamped, wait 30 minutes before investigating, **then take these steps:**

- *STEP 1.* Check the firing wire connection to the blasting machine to be sure that the contacts are good.

- *STEP 2.* Make two or three more attempts to fire the charge.

TEN-CAP BLASTING MACHINE

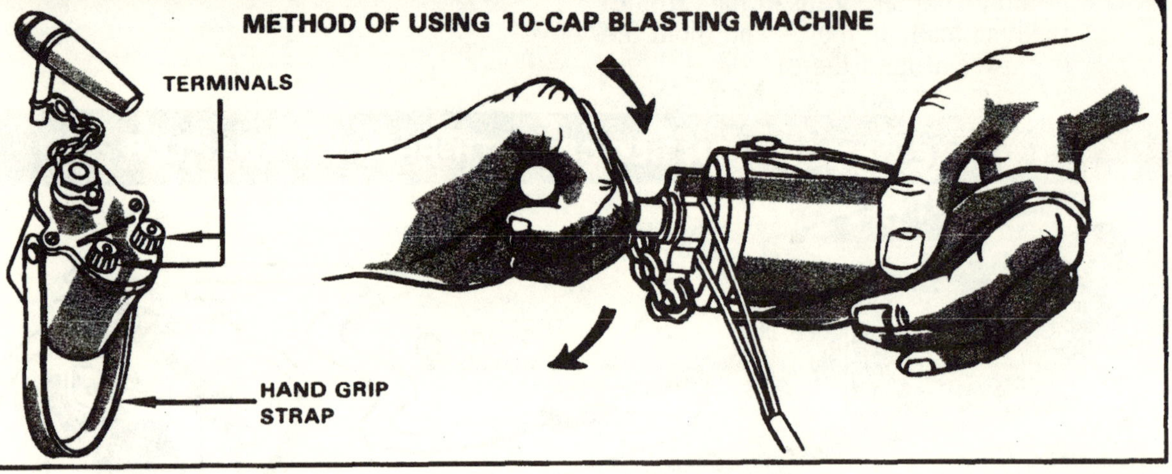

- *STEP 3.* Try to fire it again using another blasting machine.
- *STEP 4.* Disconnect the firing wire from the blasting machine and shunt (twist together) the ends of the wire.
- *STEP 5.* Move to the charge position to investigate. Take the blasting machine with you.
- *STEP 6.* Check the entire circuit, including the firing wire, for breaks and short circuits.
- *STEP 7.* Make no attempt to remove the primer or the charge.
- *STEP 8.* If the fault has not been found, place a new primed charge beside the misfire charge.
- *STEP 9.* Disconnect the old blasting cap wires from the firing wire and shunt the ends of the blasting cap wires.
- *STEP 10.* Attach the new blasting cap wires to the firing wire and fire the new charge. This should also detonate the misfire charge.

WARNING

CURRENT FROM RADIO FREQUENCY SIGNALS MAY CAUSE PREMATURE DETONATION OF ELECTRIC BLASTING CAPS. MOBILE TYPE TRANSMITTERS AND PORTABLE TRANSMITTERS ARE PROHIBITED WITHIN 50 METERS OF ANY ELECTRIC BLASTING CAPS OR ELECTRICAL FIRING SYSTEMS.

LIGHTNING IS ALSO A HAZARD TO BOTH ELECTRIC AND NONELECTRIC BLASTING CHARGES. THE ONLY SAFE PROCEDURE IS TO SUSPEND ALL BLASTING ACTIVITIES DURING AN ELECTRICAL STORM OR WHEN ONE IS THREATENING.

ELECTRIC FIRING SHOULD NOT BE CONDUCTED WITHIN 155 METERS OF ENERGIZED POWER TRANSMISSION LINES.

APPENDIX C

Obstacles

GENERAL

In combat, enemy units use obstacles to stop or slow their opponent's movement. Because of that, you may have to bypass or breach (make a gap through) those obstacles in order to continue your mission.

Two basic obstacles used by the enemy are minefield and wire obstacles. This appendix gives guidance on breaching and crossing minefield and wire obstacles.

CONTENTS

GENERAL C-1
HOW TO BREACH AND CROSS A MINEFIELD C-2
HOW TO BREACH AND CROSS WIRE OBSTACLES C-5

HOW TO BREACH AND CROSS A MINEFIELD

There are many ways to breach a minefield. One way is to probe for and mark mines to clear a footpath through the minefield.

PROBING FOR MINES

- **Remove your helmet, load-carrying equipment (LCE), watch, rings, belt, dog tags, and anything else that may hinder movement or fall off.**
- **Leave your rifle and equipment with another soldier in the team.**
- **Get a wooden stick about 30 cm (12 in) long for a probe and sharpen one of the ends. Do not use a metal probe.**
- Place the unsharpened end of the probe in the palm of one hand with your fingers extended and your thumb holding the probe.
- Probe every 5 cm (2 in) across a 1-meter front. Push the probe gently into the ground at an angle less than 45 degrees.
- Kneel (or lie down) and feel upward and forward with your free hand to find tripwires and pressure prongs before starting to probe.
- Put just enough pressure on the probe to sink it slowly into the ground. If the probe does not go into the ground.

MINE PROBE

pick or chip the dirt away with the probe and remove it by hand.

- Stop probing when a solid object is touched.
- Remove enough dirt from around the object to find out what it is.

LANES

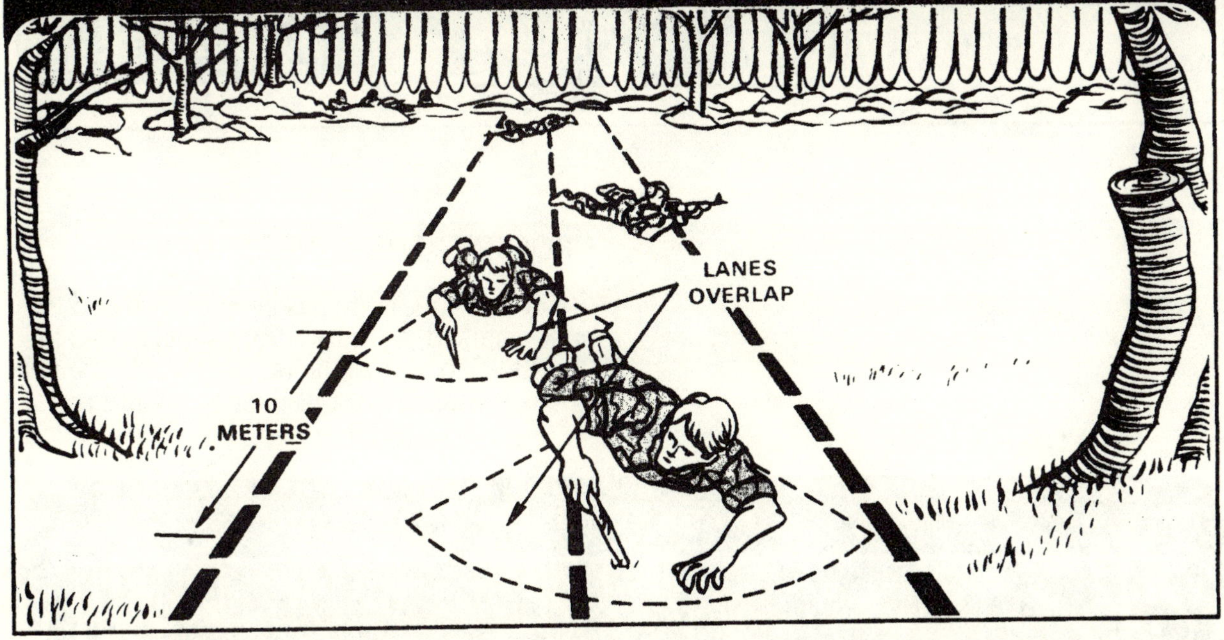

MARKING THE MINE

- Remove enough dirt from around it to see what type of mine it is.
- Mark it and report its exact location to your leader. There are several ways to mark a mine. How it is marked is not as important as having everyone understand the marking. A common way to mark a mine is to tie a piece of paper, cloth, or engineer tape to a stake and put the stake in the ground by the mine.

CROSSING THE MINEFIELD

Once a footpath has been probed and the mines marked, a security team should cross the minefield to secure the far side. After the far side is secure, the rest of the unit should cross.

HOW TO BREACH AND CROSS WIRE OBSTACLES

The enemy uses wire obstacles to separate infantry from tanks and to slow or stop infantry. His wire obstacles are similar to ours. To breach them, you should use wire cutters and bangalore torpedoes.

Breaching a wire obstacle may require stealth; for example, when done by a patrol. It may not require stealth during an attack. Breaches requiring stealth are normally done with wire cutters. Other breaches are normally done with bangalore torpedoes and wire cutters.

CUTTING THE WIRE

To cut through a wire obstacle with stealth

- **Cut only the lower strands and leave the top strand in place. That makes it less likely that the enemy will discover the gap.**

- **Cut the wire near a picket. To reduce the noise of a cut, have another soldier wrap cloth around the wire and hold the wire with both hands. Cut part of the way through the wire between the other soldier's hands and have him bend the wire back and forth until it breaks. If you are alone, wrap cloth around the wire near a picket, partially cut the wire, and then bend and break the wire.**

To breach an obstacle made of **concertina:**

- **Cut the wire and stake it back to keep the breach open.**
- **Stake the wire back far enough to allow room to crawl through or under the obstacle.**

CROSSING THE WIRE

To crawl under a **wire obstacle:**

- **Slide headfirst on your back.**
- **Push forward with your heels.**
- **Carry your weapon lengthwise on your body and steady it with one hand. To keep the wire from snagging on your clothes and equipment, let it slide along your weapon.**
- **Feel ahead with your free hand to find the next strand of wire and any trip wires or mines.**

CRAWLING UNDER A WIRE OBSTACLE

To cross over a **wire obstacle:**

- **Stay crouched down low.**
- **Feel and look for tripwires and mines.**
- **Grasp the first wire strand lightly, and cautiously lift one leg over the wire.**

- Lower your foot to the ground.
- Lift your other foot over the wire and lower it to the ground.
- Release the wire and feel for the next strand.
- To speed up a crossing, put boards or grass mats over the wire and cross on them.

USING A BANGALORE TORPEDO

A bangalore torpedo comes in a kit that has 10 torpedo sections, 10 connecting sleeves, and 1 nose sleeve. Use only the number of torpedo sections and connecting sleeves needed.

BANGALORE TORPEDO

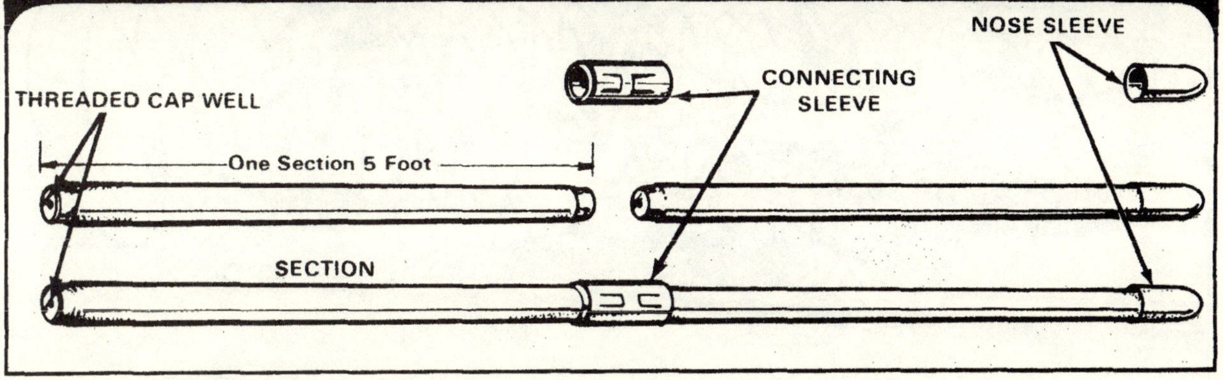

All torpedo sections have a threaded cap well at each end so that they may be assembled in any order. Use the connecting sleeves to connect the torpedo sections together. To prevent early detonation of the entire bangalore torpedo if you hit a mine while pushing it through the obstacle, attach an improvised (wooden) torpedo section to its end. That section can be made out of any wooden pole or stick that is the size of a real torpedo section. Attach the nose sleeve to the end of the wooden section.

After the bangalore torpedo has been assembled and pushed through the obstacle, prime it with either an electric or nonelectric firing system (app B).

Once the bangalore torpedo has been fired, use wire cutters to cut away any wire not cut by the explosion.

BANGALORE TORPEDO PLACEMENT

APPENDIX D

Urban Areas

GENERAL

Successful combat operations in urban areas require skills that are unique to this type of fighting. This appendix discusses some of those skills. For a more detailed discussion, see FM 90-10-1.

HOW TO MOVE

Movement in urban areas is a fundamental skill that you must master. To minimize exposure to enemy fire **while moving:**

- **Do not silhouette yourself, stay low, avoid open areas such as streets, alleys, and parks.**
- **Select your next covered position before moving.**
- **Conceal your movements by using smoke, buildings, rubble, or foliage.**
- **Move rapidly from one position to another.**
- **Do not mask your overwatching/ covering fire when you move; and stay alert and ready.**

CONTENTS

GENERAL D-1

HOW TO MOVE D-1

HOW TO ENTER A BUILDING D-7

HOW TO USE HAND GRENADES D-10

HOW TO USE FIGHTING POSITIONS D-10

HOW TO CROSS A WALL

Always cross a wall rapidly. But first, find a low spot to cross and visually reconnoiter the other side of the wall to see if it is clear of obstacles and the enemy. Next, quickly roll over the wall, keeping a low silhouette. The rapid movement and low silhouette keep the enemy from getting a good shot at you.

CROSSING A WALL

HOW TO MOVE AROUND A CORNER

Before moving around a corner, check out the area beyond it to see if it is clear of obstacles and the enemy. Do not expose yourself when checking out that area. Lie flat on the ground and do not expose your weapon beyond the corner. With your steel helmet on, look around the corner at ground level only enough to see around it. Do not expose your head any more than necessary. If there are no obstacles or enemy present, stay low and move around the corner.

CORNER MOVEMENT

HOW TO MOVE PAST A WINDOW

When moving past a window on the first floor of a building, stay below the window level. Take care not to silhouette yourself in the window, and stay close to the side of the building.

HIGH WINDOW MOVEMENT

When moving past a window in a basement, use the same basic techniques used in passing a window on the first floor. However, instead of staying below the window, step or jump over it without exposing your legs.

BASEMENT WINDOW MOVEMENT

HOW TO MOVE PARALLEL TO A BUILDING

When you must move parallel to a building, use smoke for concealment and have someone to overwatch your move. Stay close to the side of the building. Use shadows if possible, and stay low. Move quickly from covered position to covered position.

PARALLEL MOVEMENT

HOW TO CROSS OPEN AREAS

Whenever possible, you should avoid kill zones such as streets, alleys, and parks. They are natural kill zones for enemy machine guns. When you must cross an open area, do it quickly. Use the shortest route across the area. Use smoke to conceal your move and have someone overwatch you.

If you must go from point A to point C, as depicted in the illustration, do not move from point A straight to point C. This is the longest route across the open area and gives the enemy more time to track and hit you.

Instead of going from point A straight to point C, select a place (point B) to move to, using the shortest route across the open area.

Once on the other side of the open area, move to point C using the techniques already discussed.

HOW TO MOVE IN A BUILDING

When moving in a building, do not silhouette yourself in doors and windows. Move past them as discussed for outside movement.

If forced to use a hallway, do not present a large target to the enemy. Hug the wall and get out of the hallway quickly.

HOW TO ENTER A BUILDING

When entering a building, take every precaution to get into it with minimum exposure to enemy fire and observation. **Some basic rules are:**

- **Select an entry point before moving.**
- **Avoid windows and doors.**
- **Use smoke for concealment.**
- **Make new entry points by using demolitions or tank rounds.**
- **Throw a hand grenade through the entry point before entering.**
- **Quickly follow the explosion of the hand grenade.**
- **Have your buddy overwatch you as you enter the building.**
- **Enter at the highest level possible.**

HIGH LEVEL ENTRIES

The preferred way to clear a building is to clear from the top down. That is why you should enter at the highest level possible. If a defending enemy is forced down to the ground level, he may leave the building, thus exposing himself to the fires outside the building.

HIGH LEVEL ENTRY

If the enemy is forced up to the top floor, he may fight even harder than normal or escape over the roofs of other buildings.

You can use ropes, ladders, drain pipes, vines, helicopters, or the roofs and windows of adjoining buildings to reach the top floor or roof of a building. In some cases, you can climb onto another soldier's shoulders and pull yourself up. You can attach a grappling hook to one end of a rope and throw the hook to the roof, where it can snag something to hold the rope in place.

ROOF LANDING

LOW LEVEL ENTRIES

There will be times when you can't enter from an upper level or the roof. In such cases, entry at the ground floor may be your only way to get into the building. When making low level entries, avoid entries through windows and doors as much as possible. They are often booby trapped and are probably covered by enemy fire.

When making low level entries, use demolitions, artillery, tanks, antitank weapons, or similar means to make an entry point in a wall. Before entering the entry point, throw a cooked-off hand grenade through the entry point to reinforce the effects of the first blast.

LOW LEVEL ENTRY

HOW TO USE HAND GRENADES

When fighting in built-up areas, use hand grenades to clear rooms, hallways, and buildings. Throw a hand grenade before entering a door, window, room, hall, stairwell, or any other entry point. Before throwing a hand grenade, let it cook off for 2 seconds. That keeps the enemy from throwing it back before it explodes.

To cook off a hand grenade remove your thumb from the safety lever; allow the lever to rotate out and away from the grenade; then count one thousand one, one thousand two, and throw it.

The best way to put a grenade into an upper-story opening is to use a grenade launcher.

HAND GRENADE TOSS

When you throw a hand grenade into an opening, stay close to the building, using it for cover. Before you throw the hand grenade, select a safe place to move to in case the hand grenade does not go into the opening or in case the enemy throws it back. Once you throw the hand grenade, take cover. After the hand grenade explodes, move into the building quickly.

HOW TO USE FIGHTING POSITIONS

Fighting positions in urban areas are different from those in other types of terrain. They are not always prepared as discussed in **chapter 2.** In some cases, you must use hasty fighting positions which are no more than whatever cover is available.

CORNERS OF BUILDINGS

When using a corner of a building as a fighting position, you must be able to fire from either shoulder. Fire from the shoulder that lets you keep your body close to the wall of the building and expose as little of yourself as possible. If possible, fire from the prone position.

CORNER POSITION

WALLS

When firing from behind a wall, fire around it if possible, not over it. Firing around it reduces the chance of being seen by the enemy. Always stay low, close to the wall, and fire from the shoulder that lets you stay behind cover.

WALL POSITION

WINDOWS

When using a window as a fighting position, do not use a standing position, as it exposes most of your body. Standing may also silhouette you against a light-colored interior wall or a window on the other side of the building. Do not let the muzzle of your rifle extend beyond the window, as that may give away your position. The enemy may see the muzzle or the flash of the rifle.

The best way to fire from a window is to get well back into the room. That prevents the muzzle or flash from being seen. Kneel to reduce exposure.

WINDOW POSITION STANCE

To improve the cover provided by a window, barricade the window but leave a small hole to fire through. Also barricade other windows around your position. That keeps the enemy from knowing which windows are being used for fighting positions. Use boards from the interior walls of the building or any other material to barricade the window. The barricade material should be put on in an irregular pattern so that the enemy cannot determine which window is being used.

Place sandbags below and on the sides of the window to reinforce it and to add cover. Remove all the glass in the window to prevent injury from flying glass.

REINFORCED WINDOW POSITION

PEAKS OF ROOFS

A peak of a roof can provide a vantage point and cover for a fighting position. It is especially good for a sniper position. When firing from a rooftop, stay low and do not silhouette yourself.

ROOFTOP POSITION

A chimney, smokestack, or any other structure extending from a roof can provide a base behind which you can prepare a position. If possible, remove some of the roofing material so that you can stand inside the building on a beam or platform with only your head and shoulders above the roof. Use sandbags to provide extra cover.

POSITION INSIDE ROOF

If there are no structures extending from a roof, prepare the position from underneath the roof and on the enemy side. Remove enough of the roofing material to let you see and cover your sector through it. Use sandbags to add cover. Stand back from the opening and do not let the muzzle or flash of your rifle show through the hole. The only thing that should be noticeable to the enemy is the missing roofing material.

POSITION BENEATH ROOF

LOOPHOLES

A loophole blown or cut in a wall provides cover for a fighting position. Using loopholes reduces the number of windows that have to be used. Cut or blow several loopholes in a wall so the enemy cannot tell which one you are using. When using a loophole, stay back from it. Do not let the muzzle or flash of your rifle show through it.

To reinforce a loophole and add cover, put sandbags around it. If you will be firing from a prone position on the second floor, put sandbags on the floor to lie on. That will protect you from explosions on the first floor. Use a table with sandbags on it or some other sturdy structure to provide overhead cover. That will protect you from falling debris.

LOOPHOLE

REINFORCED LOOPHOLE

APPENDIX E

Tracking

GENERAL

In all operations, you must be alert for signs of enemy activity. Such signs can often alert you to an enemy's presence and give your unit time to prepare for contact. The ability to track an enemy after he has broken contact also helps you regain contact with him.

TRACKER QUALITIES

Visual tracking is following the path of men or animals by the signs they leave, primarily on the ground or vegetation. Scent tracking is following men or animals by their smell.

Tracking is a precise art. You need a lot of practice to achieve and keep a high level of tracking skill. You should be familiar with the general techniques of tracking to enable you to detect the presence of a hidden enemy and to follow him, to find and avoid mines or booby-traps, and to give early warning of ambush.

CONTENTS

GENERAL E-1
TRACKER QUALITIES E-1
FUNDAMENTALS OF TRACKING E-2
DISPLACEMENT E-2
STAINING E-6
WEATHERING E-7
LITTERING E-8
CAMOUFLAGE E-8
INTERPRETATION/IMMEDIATE USE INTELLIGENCE E-8
TRACKING TEAMS E-9
TRACKER DOGS E-10
COUNTERTRACKING E-10

With common sense and a degree of experience, you can track another person. However, **you must develop the following traits and qualities:**

- **Be patient.**
- **Be able to move slowly and quietly, yet steadily, while detecting and interpreting signs.**
- **Avoid fast movement that may cause you to overlook signs, lose the trail, or blunder into an enemy unit.**
- **Be persistent and have the skill and desire to continue the mission even though signs are scarce or weather or terrain is unfavorable.**
- **Be determined and persistent when trying to find a trail that you have lost.**
- **Be observant and try to see things that are not obvious at first glance.**
- **Use your senses of smell and hearing to supplement your sight.**
- **Develop a feel for things that do not look right. It may help you regain a lost trail or discover additional signs.**
- **Know the enemy, his habits, equipment, and capability.**

FUNDAMENTALS OF TRACKING

When tracking an enemy, you should build a picture of him in your mind. Ask yourself such questions as: How many persons am I following? How well are they trained? How are they equipped? Are they healthy? How is their morale? Do they know they are being followed?

To find the answer to such questions, use all available signs. A sign can be anything that shows you that a certain act took place at a particular place and time. For instance, a footprint tells a tracker that at a certain time a person walked on that spot.

The six fundamentals of tracking **are:**

- **Displacement.**
- **Staining.**
- **Weathering.**
- **Littering.**
- **Camouflaging.**
- **Interpretation and/or immediate use intelligence.**

Any sign that you find can be identified as one or more of the first five fundamentals. In the sixth fundamental, you combine the first five and use all of them to form a picture of the enemy.

DISPLACEMENT

Displacement takes place when something is moved from its original position. An example is a footprint in soft, moist ground. The foot of the person that left the print displaced the soil, leaving an indentation in the ground. By studying the print, you can determine many facts. For example, a print that was left by a barefoot person or a person with worn or frayed footgear indicates that he may have poor equipment.

HOW TO ANALYZE FOOTPRINTS

Footprints show **the following:**

- **The direction and rate of movement of a party.**

- The number of persons in a party.
- Whether or not heavy loads are carried.
- The sex of the members of a party.
- Whether the members of a party know **they are being followed.**

If the footprints are deep and the pace is long, the party is moving rapidly. Very long strides and deep prints, with toe prints deeper than heel prints, indicate that the party is running. If the prints are deep, short, and widely spaced, with signs of scuffing or shuffling, a heavy load is probably being carried by the person who left the prints.

You can also determine a person's sex by studying the size and position of the footprints. Women generally tend to be pigeon-toed, while men usually walk with their feet pointed straight ahead or slightly to the outside. Women's prints are usually smaller than men's, and their strides are usually shorter.

If a party knows that it is being followed, it may attempt to hide its tracks. Persons walking backward have a short, irregular stride. The prints have an unusually deep toe. The soil will be kicked in the direction of movement.

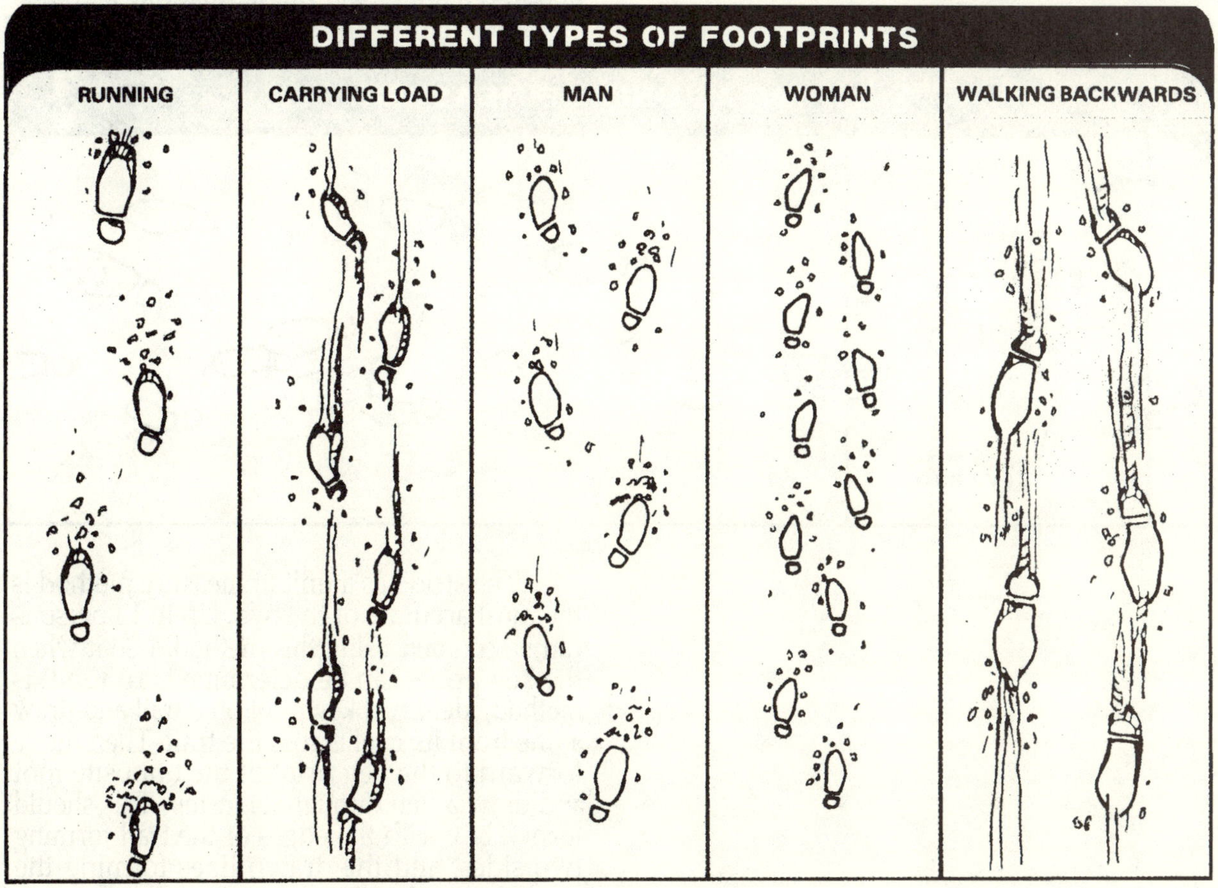

The last person walking in a group usually leaves the clearest footprints. Therefore, use his prints as the key set. Cut a stick the length of each key print and notch the stick to show the print width at the widest part of the sole. Study the angle of the key prints to determine the direction of march. Look for an identifying mark or feature on the prints, such as a worn or frayed part of the footwear. If the trail becomes vague or obliterated, or if the trail being followed merges with another, use the stick to help identify the key prints. That will help you stay on the trail of the group being followed.

Use the box method to count the number of persons in the group. There are two ways to use the box method — **the stride as a unit of measure method** and **the 36-inch box method.**

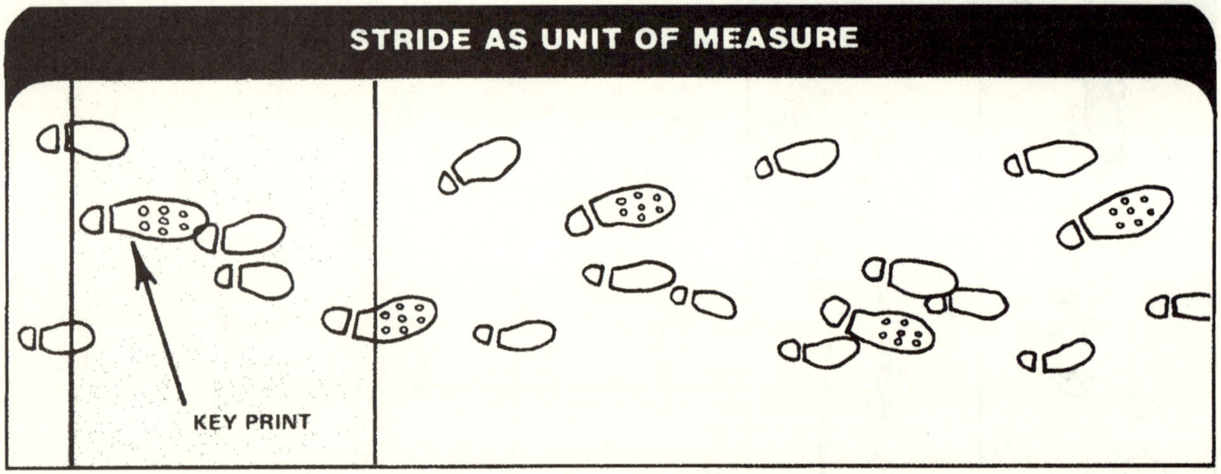

The stride as a unit of measure method is the most accurate of the two. Up to 18 persons can be counted using this method. Use it when the key prints can be determined. To use this method, identify a key print on a trail and draw a line from its heel across the trail. Then move forward to the key print of the opposite foot and draw a line through its instep. This should form a box with the edges of the trail forming two sides, and the drawn lines forming the other two sides. Next, count every print or partial print inside the box to determine the number of persons. Any person walking normally would have stepped in the box at least one time. Count the key prints as one.

To use the 36-inch box method, mark off a 30- to 36-inch cross section of a trail, count the prints in the box, then divide by two to determine the number of persons that used the trail. (Your M16 rifle is 39 inches long and may be used as a measuring device.)

36-INCH BOX METHOD

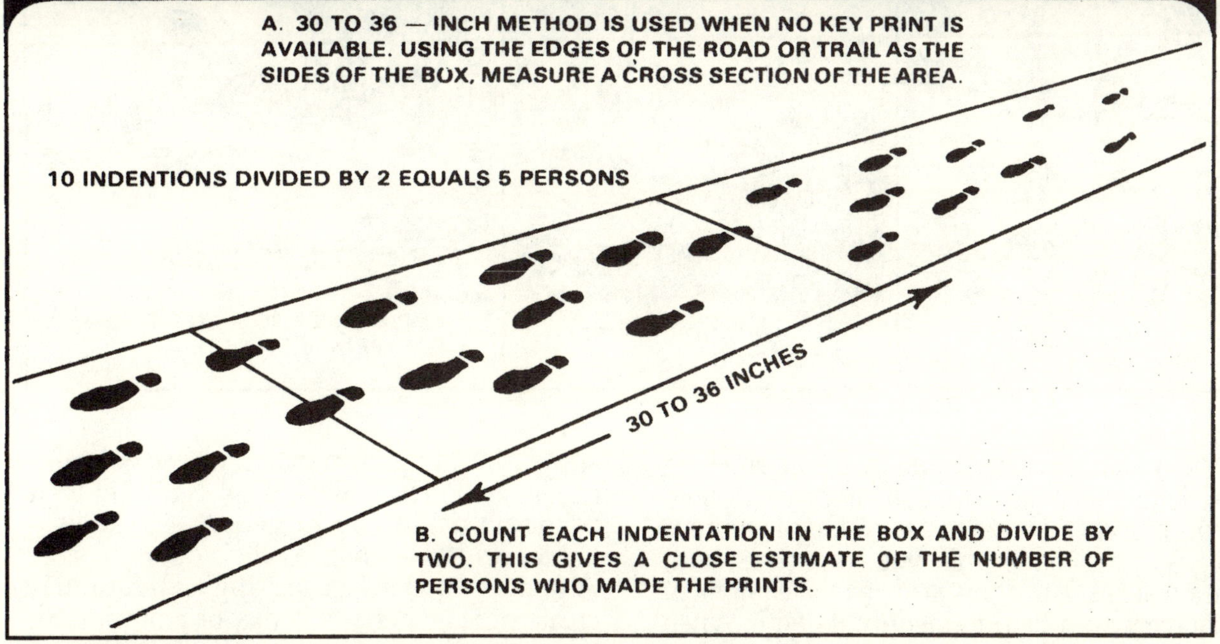

OTHER SIGNS OF DISPLACEMENT

Footprints are only one example of displacement. Displacement occurs when anything is moved from its original position. Other examples are such things as foliage, moss, vines, sticks, or rocks that are moved from their original places; dew droplets brushed from leaves; stones and sticks that are turned over and show a different color underneath; and grass or other vegetation that is bent or broken in the direction of movement.

Bits of cloth may be torn from a uniform and left on thorns, snags, or the ground, and dirt from boots may make marks on the ground.

Another example of displacement is the movement of wild animals and birds that are flushed from their natural habitats. You may hear the cries of birds that are excited by strange movements. The movement of tall grass or brush on a windless day indicates that something is moving the vegetation from its original position.

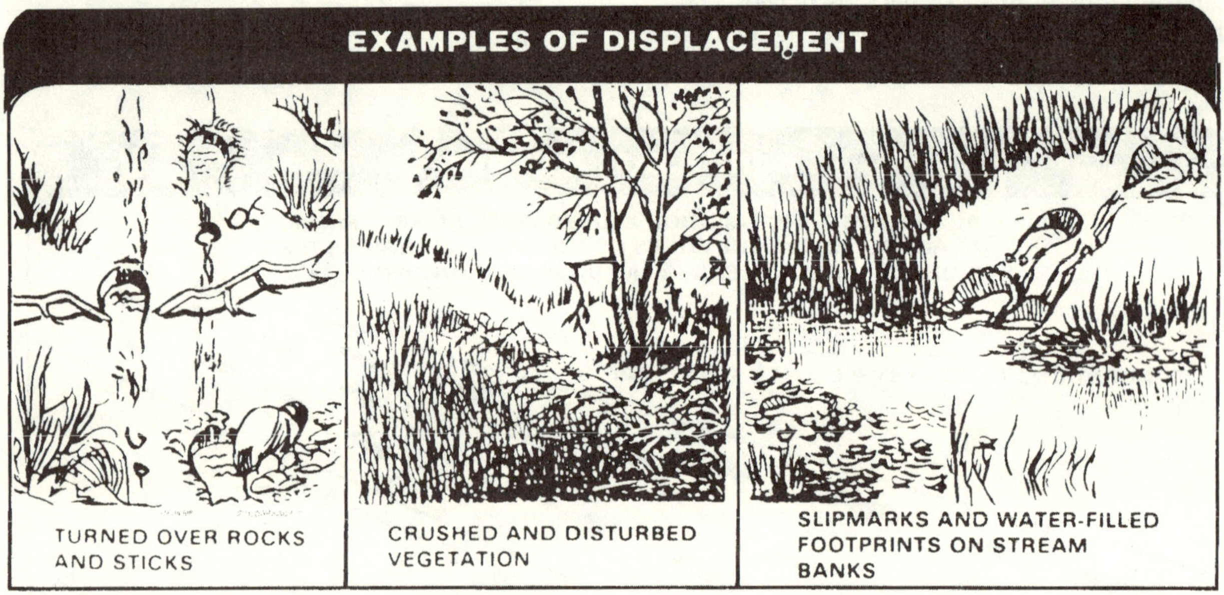

When you clear a trail by either breaking or cutting your way through heavy vegetation, you displace the vegetation. Displacement signs can be made while you stop to rest with heavy loads. The prints made by the equipment you carry can help to identify its type. When loads are set down at a rest halt or campsite, grass and twigs may be crushed. A sleeping man may also flatten the vegetation.

In most areas, there will be insects. Any changes in the normal life of these insects may be a sign that someone has recently passed through the area. Bees that are stirred up, and holes that are covered by someone moving over them, or spider webs that are torn down are good clues.

If a person uses a stream to cover his trail, algae and water plants may be displaced in slippery footing or in places where he walks carelessly. Rocks may be displaced from their original position, or turned over to show a lighter or darker color on their opposite side. A person entering or leaving a stream may create slide marks, wet banks, or footprints, or he may scuff bark off roots or sticks. Normally, a person or animal will seek the path of least resistance. Therefore, when you search a stream for exit signs, look for open places on the banks or other places where it would be easy to leave the stream.

STAINING

A good example of staining is the mark left by blood from a bleeding wound. Bloodstains often will be in the form of drops left by a wounded person. Blood signs are found on the ground and smeared on leaves or twigs.

You can determine the location of a wound on a man being followed by studying the bloodstains. If the blood seems to be dripping steadily, it probably came from a wound on his trunk. A wound in the lungs will deposit bloodstains that are pink, bubbly, frothy. A bloodstain deposited from a head wound will appear heavy, wet, and slimy, like gelatin. Abdominal wounds often mix blood with digestive juices so that the deposit will have an odor. The stains will be light in color.

Staining can also occur when a person walks over grass, stones, and shrubs with muddy boots. Thus, staining and displacement together may give evidence of movement and indicate the direction taken. Crushed leaves may stain rocky ground that is too hard for footprints.

Roots, stones, and vines may be stained by crushed leaves or berries when walked on. Yellow stains in snow may be urine marks left by personnel in the area.

In some cases, it may be hard to determine the difference between staining and displacement. Both terms can be applied to some signs. For example, water that has been muddied may indicate recent movement. The mud has been displaced and it is staining the water. Stones in streams may be stained by mud from boots. Algae can be displaced from stones in streams and can stain other stones or bark.

Water in footprints in swampy ground may be muddy if the tracks are recent. In time, however, the mud will settle and the water will clear. The clarity of the water can be used to estimate the age of the prints. Normally, the mud will clear in 1 hour. That will vary with terrain.

WEATHERING

Weather may either aid or hinder tracking. It affects signs in ways that help determine how old they are, but wind, snow, rain, and sunlight can also obliterate signs completely.

By studying the effects of weather on signs, you can determine the age of the sign. For example, when bloodstains are fresh, they may be bright red. Air and sunlight will change the appearance of blood first to a deep ruby-red color, and then to a dark brown crust when the moisture evaporates. Scuff marks on trees or bushes darken with time. Sap oozes from fresh cuts on trees but it hardens when exposed to the air.

FOOTPRINTS

Footprints are greatly affected by weather. When a foot displaces soft, moist soil to form a print, the moisture holds the edges of the print intact and sharp. As sunlight and air dry the edges of the print, small particles that were held in place by the moisture fall into the print, making the edges appear rounded. Study this process carefully to estimate the age of a print. If particles are just beginning to fall into a print, it is probably fresh. If the edges of the print are dried and crusty, the prints are probably at least an hour old. The effects of weather will vary with the terrain, so this information is furnished as a guide only.

A light rain may round out the edges of a print. Try to remember when the last rain occurred in order to put prints into a proper time frame. A heavy rain may erase all signs.

Wind also affects prints. Besides drying out a print, the wind may blow litter, sticks, or leaves into it. Try to remember the wind activity in order to help determine the age of a print. For example, you may think, "It is calm now, but the wind blew hard an hour ago. These prints have litter blown into them, so they must be over an hour old." You must be sure, however, that the litter was blown into the prints, and was not crushed into them when the prints were made.

Trails leaving streams may appear to be weathered by rain because of water running into the footprints from wet clothing or equipment. This is particularly true if a party leaves a stream in a file. From this formation, each person drips water into the prints. A wet trail slowly fading into a dry trail indicates that the trail is fresh.

WIND, SOUNDS, AND ODORS

Wind affects sounds and odors. If the wind is blowing from the direction of a trail you are following, sounds and odors are carried to you. If the wind is blowing in the same direction as the trail you are following, you must be cautious as the wind will carry your sounds toward the enemy. To find the wind direction, drop a handful of dry dirt or grass from shoulder height.

To help you decide where a sound is coming from, cup your hands behind your ears and slowly turn. When the sound is loudest, you are probably facing the origin of sound. When moving, try to keep the wind in your face.

SUN

You must also consider the effects of the sun. It is hard to look or aim directly into the sun. If possible, keep the sun at your back.

LITTERING

Poorly trained units may leave trails of litter as they move. Gum or candy wrappers, ration cans, cigarette butts, remains of fires, or human feces are unmistakable signs of recent movement.

Weather affects litter. Rain may flatten or wash litter away, or turn paper into pulp. Winds may blow litter away from its original location. Ration cans exposed to weather will rust. They first rust at the exposed edge where they were opened. Rust then moves in toward the center. Use your memory to determine the age of litter. The last rain or strong wind can be the basis of a time frame.

CAMOUFLAGE

If a party knows that you are tracking it, it will probably use camouflage to conceal its movement and to slow and confuse you. Doing so, however, will slow it down. Walking backward, brushing out trails, and moving over rocky ground or through streams are examples of camouflage that can be used to confuse you.

The party may move on hard surfaced, frequently traveled roads or try to merge with traveling civilians. Examine such routes with extreme care, because a well-defined approach that leads to the enemy will probably be mined, ambushed, or covered by snipers.

The party may try to avoid leaving a trail. Its members may wrap rags around their boots, or wear soft-soled shoes to make the edges of their footprints rounder and less distinct. The party may exit a stream in column or line to reduce the chance of leaving a well-defined exit.

If the party walks backward to leave a confusing trail, the footprints will be deepened at the toe, and the soil will be scuffed or dragged in the direction of movement.

If a trail leads across rocky or hard ground, try to work around that ground to pick up the exit trail. This process works in streams as well. On rocky ground, moss or lichens growing on the stones could be displaced by even the most careful evader. If you lose the trail, return to the last visible sign. From there, head in the direction of the party's movement. Move in ever-widening circles until you find some signs to follow.

INTERPRETATION/IMMEDIATE USE INTELLIGENCE

When reporting, do not report your interpretations as facts. Report that you have seen signs of certain things, not that those things actually exist.

Report all information quickly. The term "immediate use intelligence" includes information of the enemy that can be put to use at once to gain surprise, to keep the enemy off balance, or to keep him from escaping an area. A commander has many sources of intelligence. He puts the information from those sources together to help determine where an enemy is, what he may be planning, and where he may be going.

Information you report gives your leader definite information on which he can act at once. For example, you may report that your leader is 30 minutes behind an enemy unit, that the enemy is moving north, and that he is now at a certain place. That gives your leader information on which he can act at once. He could then have you keep on tracking and move another unit to attack the enemy. If a trail is found that has signs of recent enemy activity, your leader can set up an ambush on it.

TRACKING TEAMS

Your unit may form tracking teams. The lead team of a moving unit can be a tracking team, or a separate unit may be a tracking team. There are many ways to organize such teams, and they can be any size. There should, however, be a leader, one or more trackers, and security for the trackers. A typical organization has three trackers, three security men, and a team leader with a radiotelephone operator (RATELO).

When a team is moving, the best tracker should be in the lead, followed by his security. The two other trackers should be on the flanks, each one followed and overmatched by his security. The leader should be where he can best control the team. The RATELO should be with the leader.

TRACKING TEAM

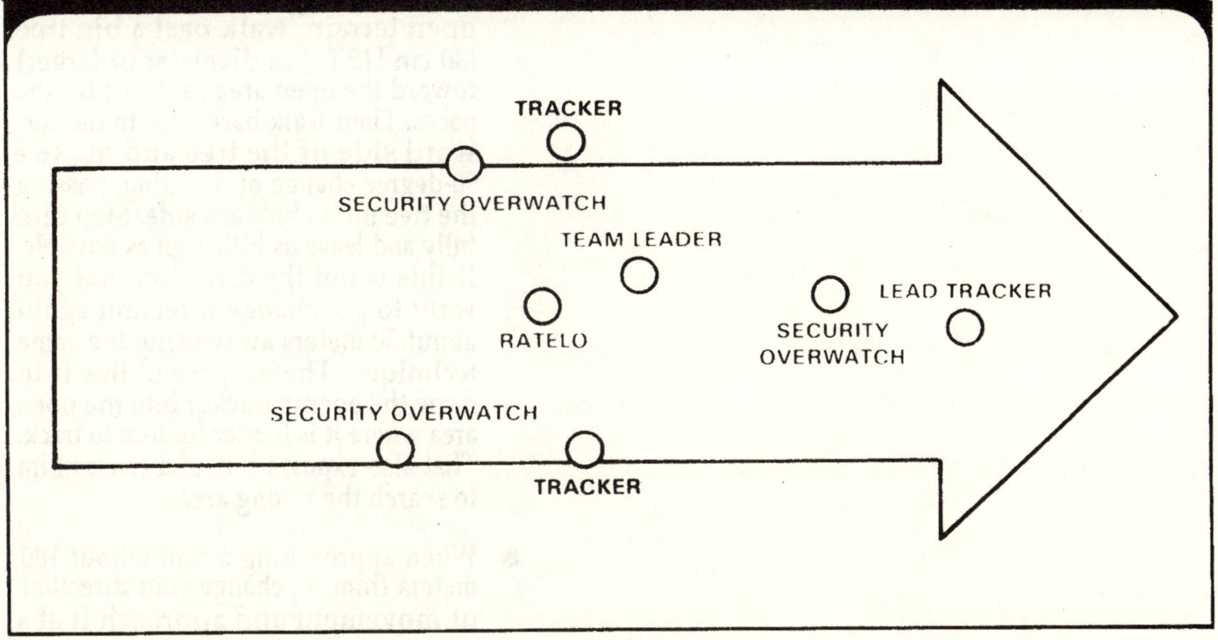

TRACKER DOGS

Tracker dogs may be used to help track an enemy. Tracker dogs are trained and used by their handlers. A dog tracks human scent and the scent of disturbed vegetation caused by man's passing.

Tracker dogs should be used with tracker teams. The team can track visually, and the dog and handler can follow. If the team loses the signs, then the dog can take over. A dog can track faster than a man tracks, and a dog can track at night.

A tracker dog is trained not to bark and give away the team. It is also trained to avoid baits, cover odors, and deodorants used to throw it off the track.

COUNTERTRACKING

In addition to knowing how to track, you must know how to counter an enemy tracker's efforts to track you. Some countertracking techniques are discussed in the following paragraphs:

- While moving from close terrain to open terrain, walk past a big tree (30 cm [12 in] in diameter or larger) toward the open area for three to five paces. Then walk backward to the forward side of the tree and make a 90-degree change of direction, passing the tree on its forward side. Step carefully and leave as little sign as possible. If this is not the direction that you want to go, change direction again about 50 meters away using the same technique. The purpose of this is to draw the enemy tracker into the open area where it is harder for him to track. That also exposes him and causes him to search the wrong area.

- When approaching a trail (about 100 meters from it), change your direction of movement and approach it at a

45-degree angle. When arriving at the trail, move along it for about 20 to 30 meters. Leave several signs of your presence. Then walk backward along the trail to the point where you joined it. At that point, cross the trail and leave no sign of your leaving it. Then move about 100 meters at an angle of 45 degrees, but this time on the other side of the trail and in the reverse of your approach. When changing direction back to your original line of march, the big tree technique can be used. The purpose of this technique is to draw the enemy tracker along the easier trail. You have, by changing direction before reaching the trail, indicated that the trail is your new line of march.

APPROACH TRAIL

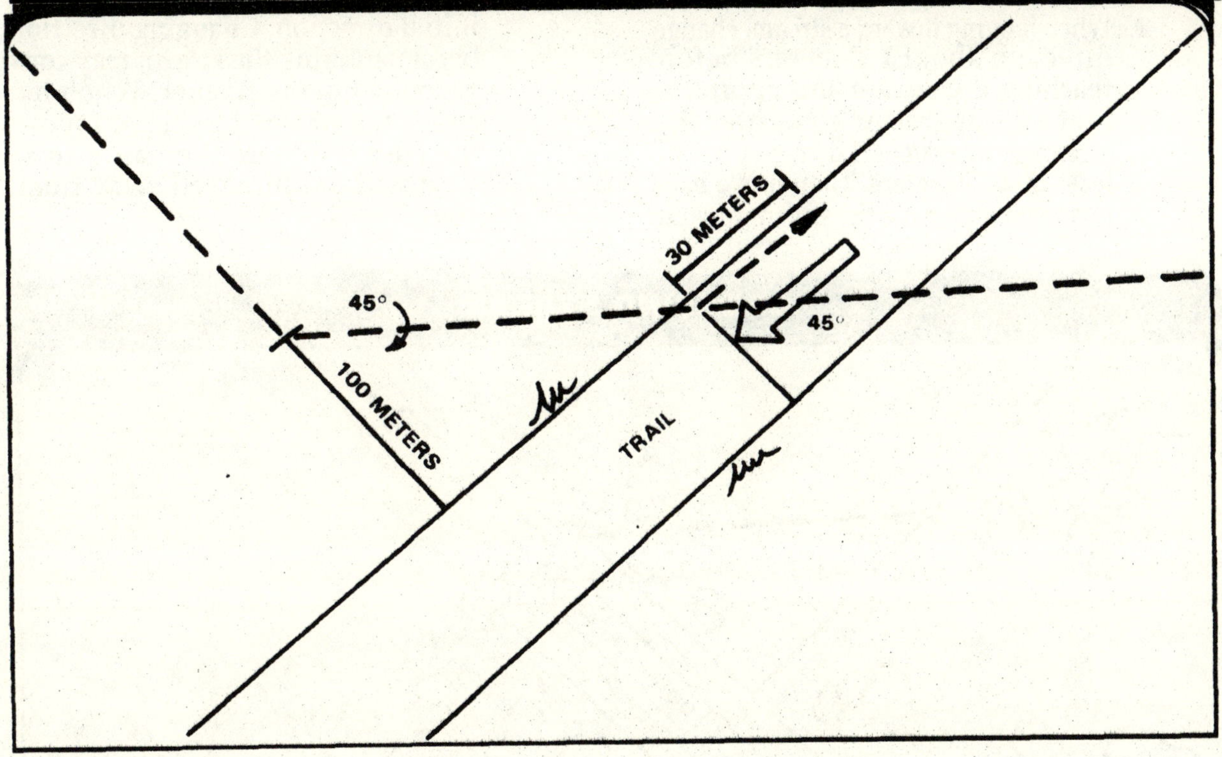

- To leave a false trail and to get an enemy tracker to look in the wrong direction, walk backward over soft ground. Continue this deception for about 20 to 30 meters or until you are on hard ground. Use this technique when leaving a stream. To further confuse the enemy tracker, use this technique several times before actually leaving the stream.

FALSE TRAIL LEAVING STREAM

- When moving toward a stream, change direction about 100 meters before reaching the stream and approach it at a 45-degree angle. Enter the stream and proceed down it for at least 20 to 30 meters. Then move back upstream and leave the stream in your initial direction. Changing direction before entering the stream may confuse the enemy tracker. When he enters the stream, he should follow the false trail until the trail is lost. That will put him well away from you.

CROSSING STREAM

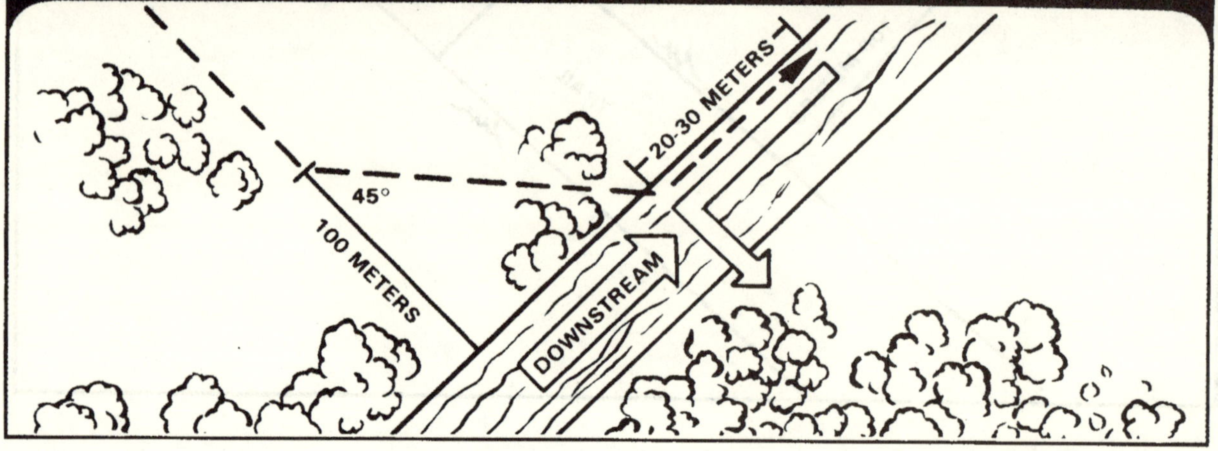

- When your direction of movement parallels a stream, use the stream to deceive an enemy tracker. Some tactics that will help elude a tracker are as follows:
  - ☐ Stay in the stream for 100 to 200 meters.
  - ☐ Stay in the center of the stream and in deep water.
  - ☐ Watch for rocks or roots near the banks that are not covered with moss or vegetation and leave the stream at that point.
  - ☐ Walk out backward on soft ground.
  - ☐ Walk up a small, vegetation-covered tributary and exit from it.
- When being tracked by an enemy tracker, the best bet is to either try to outdistance him or to double back and ambush him.

PARALLEL TRAIL

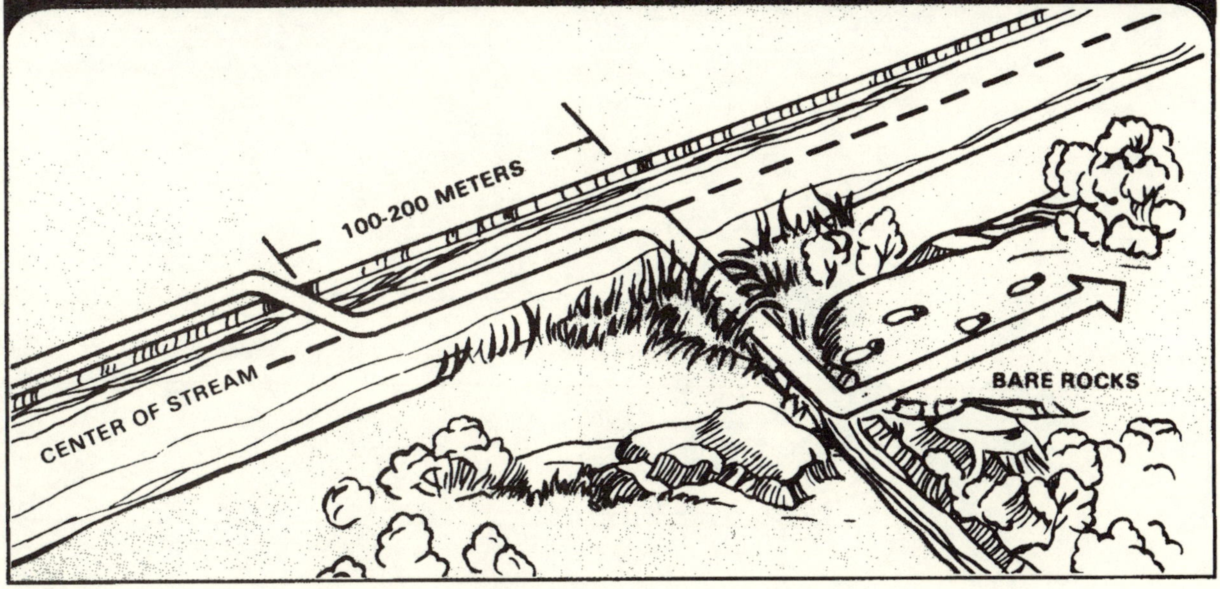

APPENDIX F

Survival, Evasion, Resistance, And Escape

GENERAL

Continuous operations and fast-moving battles increase your chances of becoming temporarily separated from your unit. Whether you are separated from a small patrol or a large unit, your mission after being separated is to **rejoin your unit.**

This appendix provides techniques to help you find your way back to your unit. For a more detailed discussion, see FM 21-76.

CONTENTS

GENERAL F-1
SURVIVAL F-1
RESISTANCE F-2
SECURITY F-4

SURVIVAL

Survival is the action of staying alive in the field with limited resources. You must try to survive when you become separated from your unit, are evading the enemy, or during the time you are a prisoner. Survival requires a knowledge of how to live off the land and take care of yourself.

EVASION

Evasion is the action you take to stay out of the hands of the enemy when separated from your unit and in an enemy area. There are several courses of action you may take to avoid capture and rejoin your unit.

You may stay in your current position and wait for friendly troops to find you. This may be a good course of action if you are sure that friendly units will continue to operate in the area, and if there are a lot of enemy units in this area.

You may break out to a friendly area. This may be a good course of action if you know where a friendly area is, and if the enemy is widely dispersed.

You may move farther into enemy territory to temporarily conduct guerrilla-type operations. This is a short-term course of action to be taken only when other courses of action are not feasible. This may be a good course of action when the enemy area is known to be lightly held, or when there is a good chance of linking up with friendly guerrillas.

You may combine two or more of these. For example, you may stay in your current position until the enemy moves out of the area and then break out to a friendly area.

There may be times when you will have to kill, stun, or capture an enemy soldier without alerting other enemy in the area. At such times, a rifle or pistol makes too much noise, and you will use a silent weapon. Some silent weapons are:

- **The bayonet.**
- **The garotte (a choke wire or cord with handles).**
- **Improvised clubs.**

In day or night, the successful use of silent weapons requires great skill and stealthy movement.

RESISTANCE

The Code of Conduct is an expression of the ideals and principles which traditionally have guided and strengthened American fighting men and the United States. It prescribes the manner in which every soldier of the United States armed forces must conduct himself when captured or when faced with the possibility of capture.

You should never surrender of your own free will. Likewise, a leader should never surrender the soldiers under his command while they still have the means to resist.

If captured, you must continue to resist in every way you can. **Some rules to follow are:**

- **Make every effort to escape and to help others escape.**
- **Do not accept special favors from the enemy.**
- **Do not give your word not to escape.**
- **Do nothing that will harm a fellow prisoner.**
- **Give no information except name, rank, social security number, and date of birth.**
- **Do not answer any questions other than those concerning your name, rank, social security number, and date of birth.**

ESCAPE

Escape is the action you take to get away

from the enemy if you are captured. The best time for you to escape is right after you are captured. You will probably be in your best physical condition at that time. Prison rations are usually barely enough to sustain life, certainly not enough to build up a reserve of energy. The physical treatment, medical care, and rations of prison life quickly cause physical weakness, night blindness, and loss of coordination and reasoning power.

The following are other reasons for **making an early escape:**

- **Friendly fire or air strikes may cause enough confusion and disorder to protide a chance of escape.**
- **The first guards you have probably will not be as well trained in handling prisoners as guards farther back.**
- **Some of the first guards may be walking wounded who are distracted by their own condition.**
- **You know something about the area where you are captured and may know the locations of nearby friendly units.**
- **The way you escape depends on what you can think of to fit the situation.**
- **The only general rules are to escape early and escape when the enemy is distracted.**

Once you have escaped, it may not be easy to contact friendly troops — even when you know where they are. You should contact a friendly unit as you would if you were a member of a lost patrol. You should time your movement so that you pass through enemy units at night and arrive at a friendly unit at dawn. A good way to make contact is to find a ditch or shallow hole to hide in where you have cover from both friendly and enemy fire. At dawn, you should attract the attention of the friendly unit by waving a white cloth, shouting, showing a panel, or some other way. This should alert the friendly unit and prepare it to accept you. After the unit has been alerted, you should shout who you are, what your situation is, and ask for permission to move toward the unit.

SECURITY

In combat, you must always think of security. You must do everything possible for the security of yourself and your unit.

The following are some basic things to do **for security:**

- **Be awake and alert.**
- **Stay dressed and ready for action.**
- **Keep your equipment packed when it is not being used.**
- **Keep your equipment and weapon in good operating condition.**
- **Use camouflage.**
- **Move around only when necessary. Stay as quiet as possible.**
- **Look and listen for enemy activity in your sector.**
- **Use lights only when necessary.**
- **Do not write information about an operation on your map.**
- **Do not take notes or papers about an operation into combat.**
- **Do not take personal items into combat.**
- **Do not leave trash lying about.**
- **Tie or tape down equipment to keep it from rattling.**
- **Use challenge and password.**
- **Do not give military information to strangers.**
- **Remember the Code of Conduct.**

APPENDIX G

Weapons And Fire Control

GENERAL

You must know how to fire your weapon and how to control your fire. This appendix covers the characteristics of the weapons you will be using and discusses characteristics of fire and methods of fire control.

CONTENTS

GENERAL G-1

WEAPONS G-2

CHARACTERISTICS OF FIRE G-11

FIRE DISTRIBUTION G-17

COVERING THE TARGET AREA G-18

FIRE CONTROL G-19

WEAPONS

M1911A1 PISTOL

This pistol fires caliber .45 rounds. It is a semiautomatic, recoil-operated magazine-fed handgun. It fires one round each time the trigger is pulled. Its magazine holds seven rounds. The top round is stripped from the magazine and chambered by the forward movement of the slide. When the last round in the magazine has been fired, the slide stays to the rear.

M1911A1 PISTOL

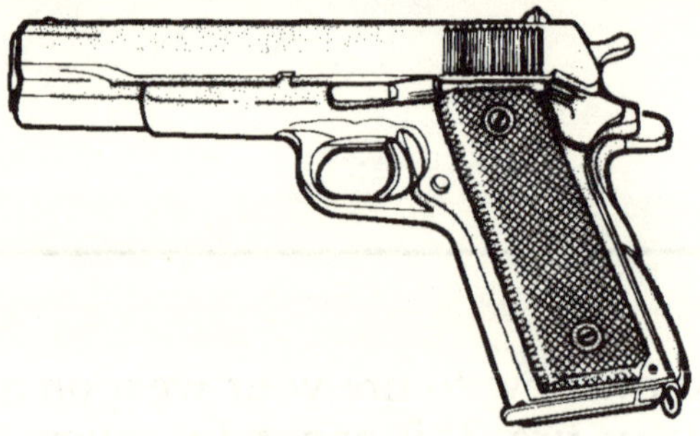

CHARACTERISTICS OF CALIBER .45 PISTOL

LENGTH	22 CM (8.6 IN)
WEIGHT (W/LOADED MAGAZINE)	1.4 KG (3 LB)
MAXIMUM RANGE	1,500 METERS
MAXIMUM EFFECTIVE RANGE	50 METERS

M16A1 RIFLE

This rifle fires 5.56-mm rounds. It is magazine-fed and gas-operated. It can shoot either semiautomatic or automatic fire through the use of a selector lever. The most stable firing positions (those which allow the most accurate fire) are the prone supported or standing supported for semiautomatic fire and the prone bipod supported for automatic fire.

M16A1 RIFLE

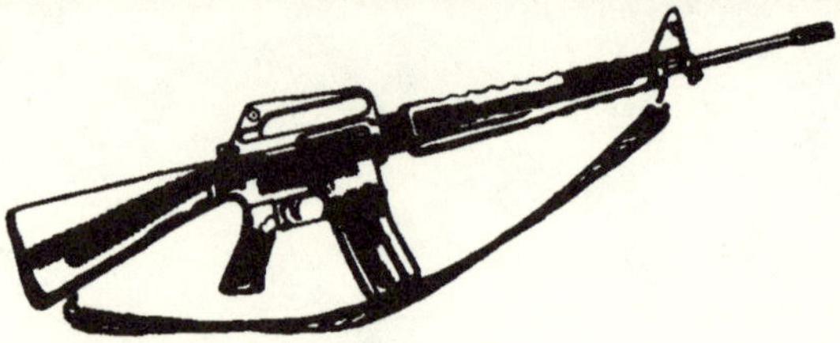

CHARACTERISTICS OF M16A1 RIFLE

WEIGHT, LOADED	
(20-RD MAGAZINE)	3.5 KG (7.6 LB)
(30-RD MAGAZINE)	3.6 KG (7.9 LB)
LENGTH (WO BAYONET)	99 CM (39 IN)
MAXIMUM RANGE	2,653 METERS
MAXIMUM EFFECTIVE RANGE	460 METERS

RATES OF FIRE

CYCLIC	700 TO 800 RD/MIN
SEMIAUTOMATIC	45 TO 65 RD/MIN
AUTOMATIC	150 TO 300 RD/MIN
SUSTAINED	12 TO 15 RD/MIN

Rate of fire is limited by a soldier's ability to aim, fire, and change magazines.

MOVING TARGET	LESS THAN 200 METERS
STATIONARY TARGET	250 METERS

M60 MACHINE GUN

This gun fires 7.62-mm rounds. It is belt-fed, gas-operated, and automatic. It has an attached bipod and a separate tripod mount. The prone position, using the M122 tripod and the traversing and elevating mechanism, allows the most accurate fire. Some vehicular mounts, such as the pedestal mount on the M151 ¼-ton vehicle, are available for this gun. When the gunner is standing, the gun may be fired from the hip, underarm, or shoulder firing position.

M60 MACHINE GUN ON MOUNT

CHARACTERISTICS OF M60 MACHINE GUN

WEIGHT,	
MACHINE GUN ONLY	10.4 KG (23 LB)
TRIPOD, W/TRAVERSING AND ELEVATING MECHANISM	8.8 KG (19.5 LB)
300 ROUNDS OF AMMUNITION	9.5 KG (21 LB)
LENGTH	110.5 CM (43.5)
MAXIMUM RANGE	3,725 METERS
TRACER BURNOUT	900 METERS
MAXIMUM RANGE OF GRAZING FIRE	600 METERS
MAXIMUM EFFECTIVE RANGE	1,100 METERS

RATES OF FIRE

CYCLIC	550 RD/MIN
SUSTAINED	100 RD/MIN
RAPID	200 RD/MIN

TYPES OF AMMUNITION

BALL
TRACER
ARMOR-PIERCING

Ranges at which a 50-50 chance of a target hit can be expected (shooting bursts of six to nine rounds):

MOVING TARGET* (BIPOD)	200 METERS
POINT TARGET* (BIPOD OR TRIPOD)	600 METERS
AREA TARGET* (BIPOD)	800 METERS
AREA TARGET** (TRIPOD)	1,100 METERS

*A point target is an area the size of a standing man.

**An area target is an area of the size that a fire team would occupy.

BIPOD MOUNTED

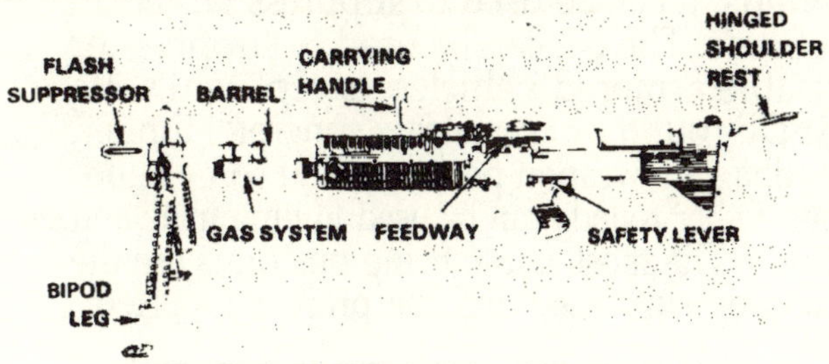

TRIPOD MOUNTED

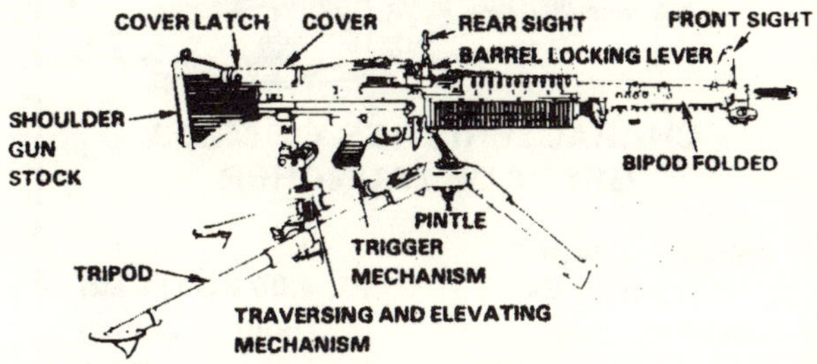

PEDESTAL MOUNTED

40-MM GRENADE LAUNCHER, M203

This grenade launcher (GL) is attached to an M16A1 rifle. The rifle has already been discussed. The GL is a single-shot, breech-loaded, pump-action weapon. It fires a variety of rounds. It can be used to suppress targets in defilade. The GL can be used to suppress or disable armored vehicles, except tanks. Its HEDP round can penetrate concrete, timber, sandbagged weapon positions, and some buildings. Other rounds can be used to illuminate and signal. The most stable firing positions are the standing supported and the prone supported.

M203 GRENADE LAUNCHER

CHARACTERISTICS OF M203 GRENADE LAUNCHER

WEIGHT, LOADED (RIFLE AND GL)	4.98 KG (11 LB)
LENGTH	99 CM (39 IN)
MAXIMUM RANGE	400 METERS
MINIMUM SAFE FIRING RANGES (HE AND TP)	
TRAINING	80 METERS
COMBAT	31 METERS
MINIMUM ARMING RANGE	14 TO 38 METERS*

Ranges at which a 50-50 chance of a target hit can be expected:

AREA TARGET (FIRE-TEAM SIZE)	350 METERS
POINT TARGET: VEHICLE OR WEAPON POSITION	200 METERS
WEAPON OPENING	125 METERS
BUNKER APERTURE	50 METERS

***This must be considered in close-in firing, as in towns and other restrictive terrain, to insure that the round will explode.**

M433 High-Explosive Dual Purpose (HEDP) Round. This round can penetrate 5 cm (2 in) of armorplate, 30 cm (12 in) of pine logs, 40 cm (16 in) of concrete blocks, or 50 cm (20 in) of sandbags at ranges up to 400 meters. It has a 5-meter casualty radius against exposed troops.

M433 HIGH-EXPLOSIVE DUAL PURPOSE (HEDP)

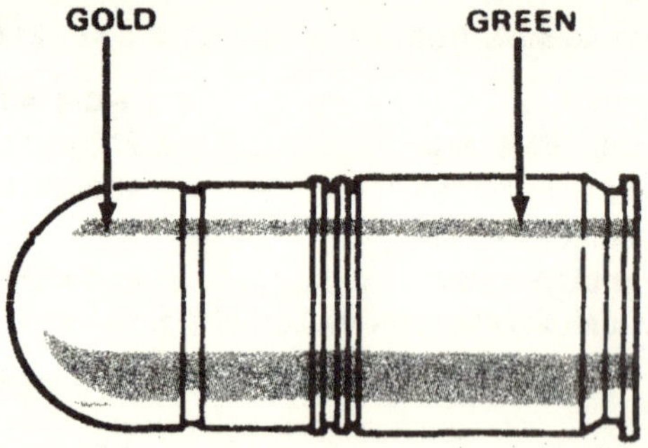

M651 CS Round. This chemical round is used to drive the enemy from bunkers or enclosed positions in built-up areas.

CTG XM651E1

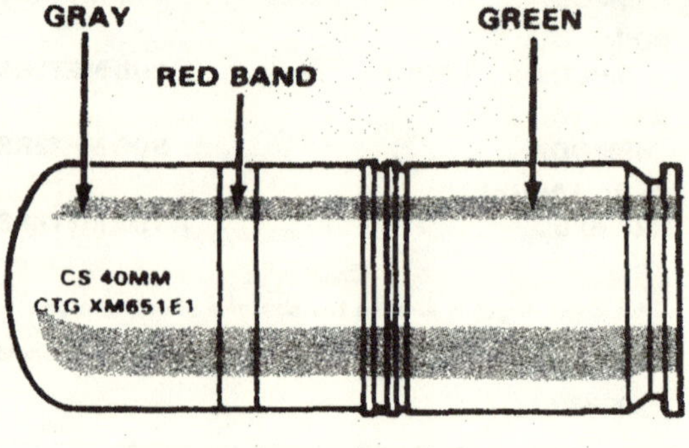

M583 White Star Parachute/M661 Green Star Parachute/M662 Red Star Parachute Rounds. These are used to signal and illuminate. One can be placed 300 meters forward of a squad to illuminate an area 200 meters in diameter for 40 seconds.

M583/M661/M662 STAR PARACHUTE

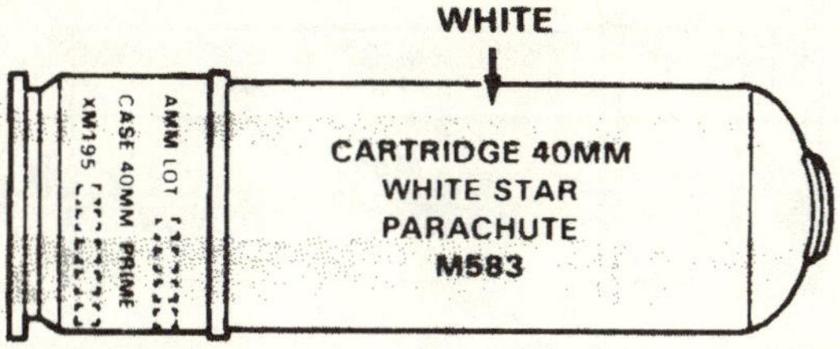

M585 White Star Cluster/M663 Green Star Cluster/M664 Red Star Cluster Rounds. They are used to signal.

NOTE: THE GREEN STAR CLUSTER MAY APPEAR WHITE IN BRIGHT SUNLIGHT.

M585/M663/M664 STAR CLUSTER

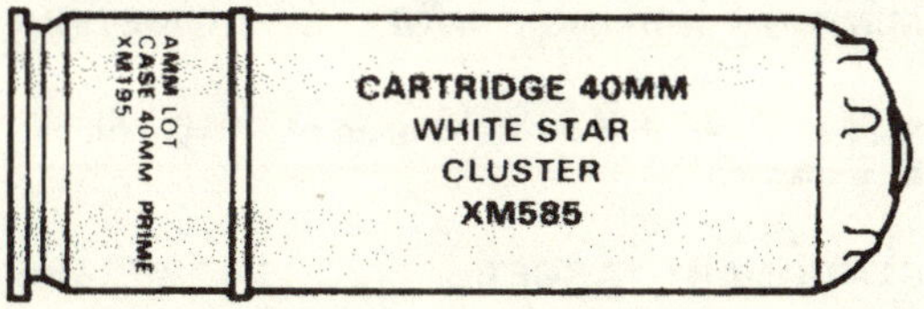

M713 Red Ground Smoke/M715 Green Smoke/ M716 Yellow Smoke Rounds. These are used to mark locations, not for screening.

M713/M715/M716 GROUND SMOKE

LIGHT ANTITANK WEAPON (LAW)

This is a shoulder-fired, short-range antitank weapon. The most stable firing positions

M72A2 LAW, 66-MM HEAT ROCKET

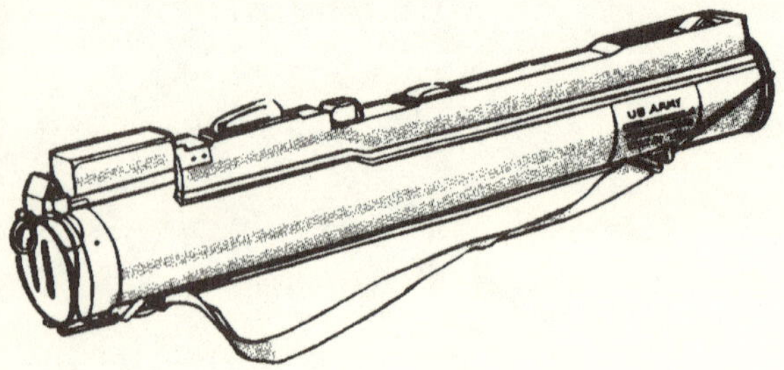

CHARACTERISTICS OF M72A2 LAW

LENGTH (CLOSED) .... 66 CM (26 IN)
LENGTH (EXTENDED) .... 89 CM (35 IN)
MAXIMUM RANGE .... 1,000 METERS
MINIMUM ARMING RANGE ...... 10 METERS

Ranges at which a 50-50 chance of a target hit can be expected.

STATIONARY TARGET .... 200 METERS
MOVING TARGET .... 165 METERS

for firing LAWs are the standing supported, prone, and prone supported.

The M72A2 LAW consists of a 66-mm HEAT (high explosive antitank) rocket in a disposable fiberglass and aluminum launcher tube. Its light weight and its ability to penetrate more than 30 cm (12 in) of armor make it useful against enemy armor, bunkers, and other hard targets out to a range of 200 meters.

The four methods of engagement with a LAW are single, sequence, pair, and volley firing. The two best methods of engagement are volley firing and pair firing.

Single firing. In single firing, you fire at a target with only one LAW. This method is used only at ranges of 50 meters or less. Beyond that range, single firing is ineffective, as the chance of a first-round hit is low.

Sequence firing. In sequence firing, you prepare several launchers for firing. After firing the first LAW, note its impact. If you get a hit, continue to fire, using the same sight picture, until the target is destroyed. If the first round is a miss, adjust the range and lead of succeeding rounds until you get a hit. Then continue to fire until the target is destroyed.

Pair firing. In pair firing, you and another gunner prepare two or more LAWs each, and fire at a target one at a time. You swap information when firing at the target. The gunner seeing a target identifies it and gives the estimated range and lead he will use (for example, TANK, 150 METERS, FAST TARGET), then fires. If the first gunner misses, the second gunner quickly announces a revised estimate of range and lead (as appropriate) and fires. Both gunners continue exchanging range and lead information until one gets a hit. Once the range and lead are determined, gunners fire at the target until it is destroyed. Pair firing is preferred over sequence firing, as it lets the gunners get hits fasten the gunner firing the second round can be ready to fire as soon as the first round impacts. In sequence firing, you must get another LAW, establish a sight picture, and fire. Pair firing also has the advantagc of having two gunners track the target at one time.

Volley firing. In volley firing, you and one or more other gunners fire at once. Before firing, each gunner prepares one or more LAWs Gunners fire on command or on signal until the target is destroyed for example, TANK, 100 METERS, SLOW TARGET, VOLLEY FIRE, READY, AIM, FIRE. Volley fire is used only when the range to the target and the lead have been determined. Range can be determined by map, by pacing, or by the results of pair firing after a target has been hit. The volley method is best because more rounds are fired at a target at one time. That increases the chance of a hit.

M202A1, MULTISHOT ROCKET LAUNCHER 66-MM (FLASH)

This is a lightweight, four-tube, 66-mm rocket launcher (RL). Aim and fire it from the right shoulder in the standing, kneeling, or prone position. It can fire a single rocket or up to four rockets semiautomatically at a rate of one rocket per second. It is reloaded with a new clip of four rockets. The brilliant splash of the bursting incendiary warhead makes it a good weapon to suppress enemy rocket gunners. When it impacts near enemy vehicles, it will make them button up. The most stable position for firing the FLASH is the standing supported position. When you fire it from a fighting position, there are two limitations. First, overhead cover can limit the elevation of the RL and therefore the range. Second, when elevating the RL, you must make sure that the rear of the launcher is outside the hole so that its backblast is not deflected on you.

M202A1 MULTISHOT ROCKET LAUNCHER (FLASH)

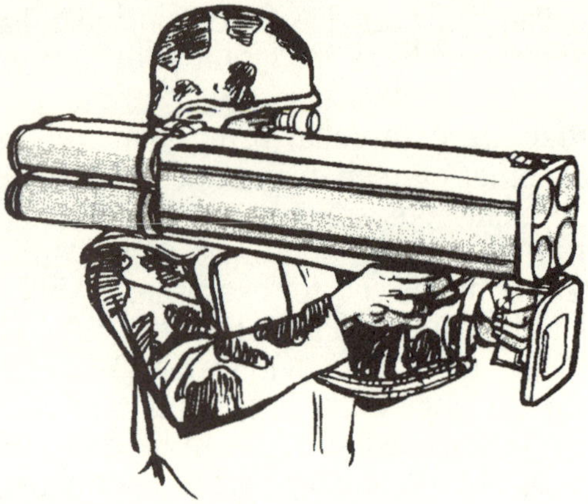

CHARACTERISTICS OF M202A1

WEIGHT, LOADED	12.1 KG (26.6 LB)
LENGTH, CLOSED	68.5 CM (27 IN)
LENGTH, EXTENDED	88.9 CM (35 IN)
MINIMUM ARMING RANGE	6 TO 13 METERS
BURSTING RADIUS OF ROCKET WARHEAD	20 METERS

Ranges at which a 50-50 chance of a target hit can be expected.

AREA TARGET (FIRE TEAM SIZE)	750 METERS
POINT TARGET	200 METERS

M47 DRAGON MEDIUM ANTITANK WEAPON

This is a wire-guided missile system. It is man-portable and shoulder-fired. The Dragon actually rests on your shoulder and the front bipod legs. It has two major components, the tracker and the round, The round (the expendable part of the system) has two major parts, the launcher and the missile. These are packaged together for handling and shipping. The launcher is both the handling and carrying container and the tube from which the missile is fired. The tracker is the reusable part of the system. It is designed for fast, easy detachment from the round.

To fire the Dragon, look through the sight in the tracker, put the crosshairs on the target, and fire. Keep the crosshairs on the target throughout the missile's flight. The missile is continuously guided along your line-of -sight. The tracker detects deviations from the line-of-sight and sends corrections to the missile by a wire link.

M47 DRAGON

CHARACTERISTICS OF DRAGON

WEIGHT (TOTAL) W/DAY TRACKER	15.30 KG (31.87 LB)
WEIGHT (TOTAL) W/NIGHT TRACKER	20.76 KG (45.89 LB)
WEIGHT OF DAY TRACKER	3.9 KG (6.58 LB)
WEIGHT OF NIGHT TRACKER	9.36 KG (20.6 LB)
WEIGHT OF ROUND	11.4 KG (25.29 LB)
LENGTH	115.4 CM (45.5 IN)
DIAMETER	29.2 CM (11.5 IN)
MINIMUM RANGE	65 METERS
MAXIMUM RANGE	1,000 METERS

CALIBER .50 MACHINE GUN

This gun is belt-fed and recoil-operated. You can fire a single shot and automatic from the M3 tripod mount or the M63 antiaircraft mount. Fire bursts of 9 to 15 rounds to hit ground targets from a stationary position. To fire at aircraft, use a continuous burst, rather than several short bursts. While firing on the move, "walk" long bursts into the target. You can suppress enemy antitank guided missile (ATGM) gunners, vehicles, and troops with a heavy volume of fire from the caliber .50 machine gun until a friendly maneuver element can destroy or bypass the enemy.

CALIBER .50 MACHINE GUN

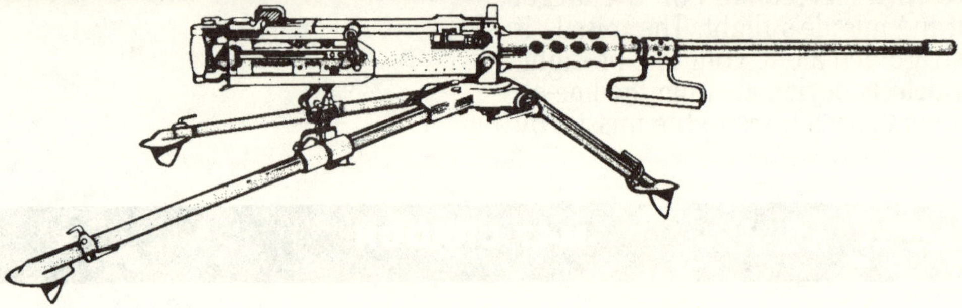

CHARACTERISTICS OF CALIBER .50

WEIGHT OF MACHINE GUN . . . . 38 KG (84 LB)

WEIGHT OF TRIPOD . . . . 20 KG (44 LB)

LENGTH OF MACHINE GUN . . . . 165 CM (65 IN)

MAXIMUM RANGE OF GRAZING FIRE . . . . 800 METERS

TRACER BURNOUT . . . . 2,200 METERS

MAXIMUM RANGE . . . . 6,800 METERS

RATES OF FIRE

SUSTAINED . . . . 40 OR LESS RD/MIN

RAPID . . . . MORE THAN 40 RD/MIN

CYCLIC . . . . 450 TO 550 RD/MIN

TYPES OF AMMUNITION

BALL

TRACER

ARMOR-PIERCING

ARMOR-PIERCING INCENDIARY

Ranges at which a 50-50 chance of a target hit can be expected:

(Tripod mounted, firing bursts of 9 to 15 rd)

POINT TARGET (MAN) . . . . 700 METERS

POINT TARGET (VEHICLE) . . . . 1,100 METERS

AREA TARGET . . . . 1,600 METERS

(Cupola mounted, stationary vehicle, firing bursts of 9 to 15 rd)

POINT TARGET (MAN) . . . . 500 METERS

POINT TARGET (VEHICLE) . . . . 600 METERS

AREA TARGET . . . . 1,000 METERS

(Cupola mounted, moving vehicle, firing bursts of 15 to 30 rd)

AREA TARGET . . . . 300 METERS

Range at which a 30% probability of hit can be expected:

SQUAD-SIZE PLATOON . . . . 500 METERS

M67 90-MM RECOILLESS RIFLE (RCLR)

This RCLR is a breech-loaded, single-shot, man-portable, crew-served weapon. You can use it in both antitank and antipersonnel roles. You can fire it from the ground, using the bipod or the monopod, or from the shoulder. The most stable firing position is the prone position.

M67 90-MM RECOILLESS RIFLE

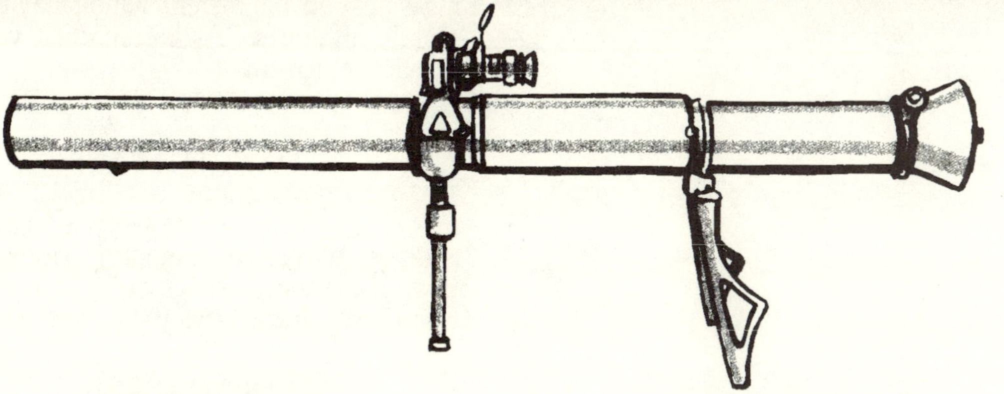

CHARACTERISTICS OF 90-MM RCLR

WEIGHT (COMPLETE, WITH SIGHT)	17.5 KG (37.5 LB)
LENGTH	135 CM (53 IN)
MAXIMUM RANGE (APPROX)	2,100 METERS
ARMING RANGE	30 TO 35 METERS

TYPES OF AMMUNITION

HEAT

TARGET PRACTICE

CANISTER (ANTIPERSONNEL)

Ranges at which a 50-50 chance of a target hit can be expected:

STATIONARY TARGET	300 METERS
MOVING TARGET	200 METERS

CHARACTERISTICS OF FIRE TRAJECTORY

This is the path of a projectile from a weapon to the point of impact.

At ranges out to 300 meters, the trajectory of rifle fire is almost flat. For greater ranges, you must raise the rifle muzzle, thus raising the height of the trajectory.

The GL has a high trajectory that is different from that of a rifle. The GL muzzle velocity is slow when compared to that of a rifle, but it is fast enough to have a flat trajectory out to 150 meters. For targets at greater ranges (150 to 350 meters), you must hold the GL about 20 degrees above the horizontal. This results in a higher trajectory and increases the time of flight of the grenade to its target. Because the trajectory is high and the time of flight long at ranges beyond 150 meters, winds may blow the grenade off course. As a grenadier, you must compensate for this.

DANGER SPACE

This is the space between a weapon and its target where the trajectory does not rise above the average height of a standing man (1.8 meters). It includes the beaten zone.

DEAD SPACE

Any area within a weapon's sector that cannot be hit by fire from that weapon is dead space.

DEAD SPACE

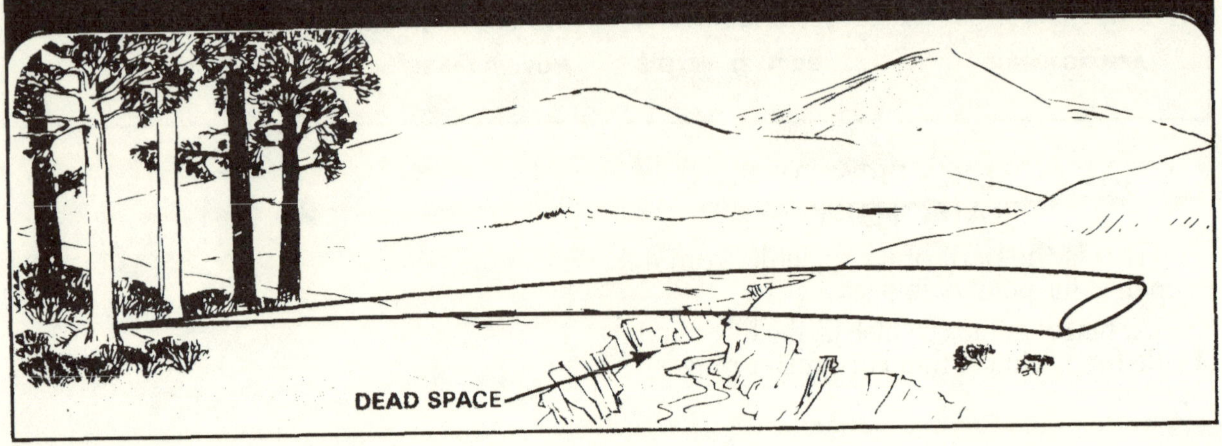

CONE OF FIRE

This is the cone-shaped pattern formed by the paths of rounds in a group or burst. The paths of the rounds differ and form a cone because of gun vibration, wind changes, and variations in ammunition.

BEATEN ZONE

The area on the ground where the rounds in a cone of fire fall is the beaten zone.

CASUALTY RADIUS

This is the area around a projectile's point of impact in which soldiers could be killed or injured by either the concussion or fragmentation of the projectile.

CLASSES OF FIRE

Fire is classified with respect to the ground and the target.

Fire with respect to the **ground is:**

- **Grazing fire when most of the rounds do not rise above 1 meter from the ground.**

GRAZING FIRE

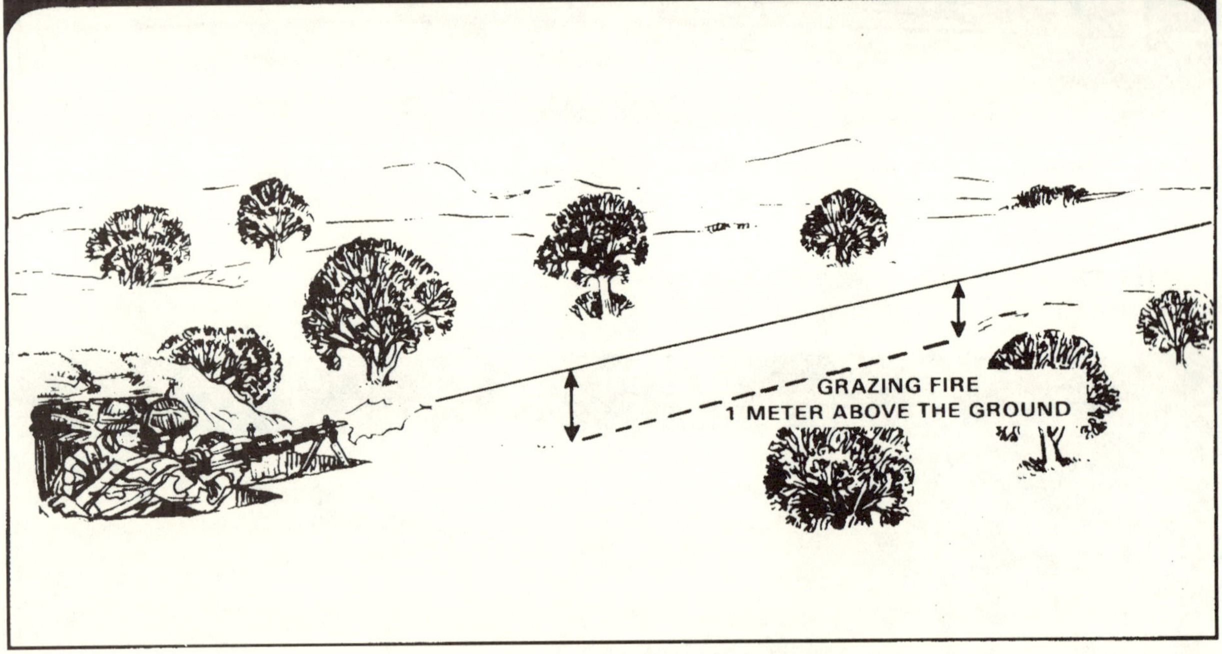

- **Plunging fire when the path of the rounds is higher than a standing man except in its beaten zone. Plunging fire is attained when firing at long ranges, when firing from high ground to low ground, and when firing into a hillside.**

PLUNGING FIRE

Fire with respect to the **target is:**

- **Frontal fire when the rounds are fired directly at the front of the target.**
- **Flanking fire when the rounds are fired at the flank of the target.**

FRONTAL FIRE

FLANKING FIRE

- Oblique fire when the long axis of the beaten zone is oblique to the long axis of the target.

- Enfilade fire when the long axis of the beaten zone is the same as the long axis of the target. It can be either frontal, flanking, or oblique. It is the best type of fire with respect to the target because it makes the best use of the beaten zone.

SUPPRESSIVE FIRE

Fire directed at the enemy to keep him from seeing, tracking, or firing at the target is suppressive fire. It can be direct or indirect fire. Smoke placed on the enemy to keep him from seeing targets is also suppressive fire.

FIRE DISTRIBUTION

When firing at an enemy position, your leader will distribute his unit's fire to cover the position. There are two ways to distribute fire on a target — point fire and area fire.

METHODS OF DISTRIBUTION

Point Fire. This is fire directed at one point; for example, an entire team firing at one bunker.

POINT FIRE

Area Fire. This is fire directed to cover an area both laterally and in depth. If your leader wants fire on a woodline, he may first fire tracers to mark its center. Then, he may have the men on his left fire to the left of the tracers and those on his right fire to the right of the tracers. This is the best and quickest way to hit all parts of an area target.

In area fire, you will fire at likely enemy positions rather than a general area. Fire first at that part of the target relative to your position in the team. Then distribute your fire over an area a few meters to the right and left of your first shot.

COVERING THE TARGET AREA

AUTOMATIC RIFLEMAN

The part of the target which you, as an automatic rifleman, can hit depends on your position and the range to the target. When possible, you cover the entire target. When firing automatic fire, you tend to fire high; so fire low at first and then work up to the target.

MACHINE GUNNER

As a machine gunner, fire into the part of the target assigned to you by your leader.

DRAGON GUNNER

As a Dragon gunner, fire into the part of the target assigned to you by your leader. Fire only at targets such as armored vehicles and key weapons. If there are no Dragon targets, fire your rifle.

GRENADIER

As a grenadier, fire your first grenade into the center of the target. Then distribute your shots over the remaining target area.

FIRE CONTROL

WAYS TO COMMUNICATE FIRE CONTROL

Your leader will control your fire. The noise and confusion of battle will limit the use of some methods of control, so he will use the way or combination of ways that does the job.

Sound. This includes both voice and devices such as whistles and horns. Sound signals are good only for short distances. Their range and reliability y are reduced by battle noise, weather, terrain, and vegetation. Voice communications may come directly from your leader to you or they may be passed from soldier to soldier.

Prearranged fire. In prearranged fire, your leader tells you to start firing once the enemy reaches a certain point or terrain feature. When using prearranged fire, you do not have to wait for an order to start firing.

Prearranged signals. In this method, your leader gives a prearranged signal when he wants you to start firing. This can be either a visual signal or a sound signal. Start firing immediately when you get the signal.

Soldier-initiated fire. This is used when there is no time to wait for orders from your leader.

Standing operating procedures (SOP). These can reduce the number of oral orders needed to control fire. SOPs must be known and understood by all members of the unit. Three SOPs are the search-fire-check SOP, the return-fire SOP, and the rate-of-fire SOP. A procedure for giving fire commands for direct fire weapons should also be SOP.

The search-fire-check SOP, follows these steps:

Step 1

- Search your assigned sectors for enemy targets.

Step 2

- Fire at any targets (appropriate for your weapon) seen in your sectors.

Step 3

- While firing in your sectors, visually check with your leader for specific orders.

The return-fire SOP tells each soldier in a unit what to do in case the unit makes unexpected contact with the enemy (in an ambush, for example). These instructions will vary from unit to unit and from position to position within those units.

The rate-of-fire SOP tells each soldier how fast to fire at the enemy. The rate of fire varies among weapons, but the principle is to fire at a maximum rate when first engaging a target and then slow the rate to a point that will keep the target suppressed. That helps keep weapons from running out of ammunition too fast.

FIRE COMMANDS

To help identify a target for a direct fire weapon and to control that weapon's fire, a leader may give a fire command to that weapon.

A fire command has the **following six parts:**

1. **Alert.**
2. **Direction.**
3. **Target Description.**
4. **Range.**
5. **Method of Fire.**
6. **Command to Fire.**

Alert. This gets your attention. The leader may alert you by calling your name or unit designation, by giving some type of visual or sound signal, by personal contact, or by any other practical way.

Direction. This tells you which way to look to see the target. The following are ways to give the direction to the target:

- **Your leader may point to a target with his arm or rifle. This will give you the general direction of the target.**
- **Your leader may fire tracer ammunition at a target to quickly and accurately identify it. However, before firing, he should show you the general direction.**
- **Your leader may designate certain features as TRPs before contact is made with the enemy. Each TRP will have a number to identify it. He may give a target's direction in relationship to a TRP. For example FROM TRP 13, RIGHT 50. That means that the target is 50 meters to the right of TRP 13.**

Target Description. This tells you what the target is. Your leader should describe it briefly, but accurately. For example MACHINE GUN POSITION IN THE WOODLINE.

Range. This tells you how far away the target is. The range is given in meters.

Method of Fire. This tells you who is to fire. It may also tell you how much ammunition to fire. For example, your leader may want only the grenadier to fire at a target. He may also want him to fire only three rounds. For example, he would say: GRENADIER, THREE ROUNDS.

Command to Fire. This tells you when to fire. It may be an oral command, or a sound or visual signal. If your leader wants to control the exact moment of fire, he may say AT MY COMMAND, (then pause until he is ready) FIRE. If he wants your fire to start upon completion of the fire command, he will simply say FIRE (without pausing).

EXAMPLE OF ORAL FIRE COMMAND

COMMAND	EXAMPLE
1. ALERT	GRENADIER
2. DIRECTION	STRAIGHT AHEAD
3. TARGET DESCRIPTION	SOLDIERS MOVING IN DITCH
4. RANGE	ONE HUNDRED
5. METHOD OF FIRE	FOUR ROUNDS
6. COMMAND TO FIRE	FIRE

Visual signals are the most common means of giving fire commands. Arm-and-hand signals, personal examples, and pyrotechnics are some of the things your leader may use for visual signals.

Your leader may use arm-and-hand signals to give fire commands when you can see him.

He may use flares and smoke grenades to mark targets in most conditions of visibility.

Your leader may use his weapon to fire on a target as a signal; you fire when he fires. Watch your leader and do as he does. He may use tracers to point out targets.

APPENDIX H

Held Expedient Antiarmor Devices

GENERAL

There are many weapons that you can use to destroy a tank or an armored personnel carrier. The weapons most frequently used are LAWs, Dragons, TOWs, mines, and high-explosive dual-purpose (HEDP) rounds of the M203 grenade launcher. There may be times, however, when you will not have these weapons available. In such cases, you may have to use field expedient devices. This appendix describes some appropriate devices.

CONTENTS

GENERAL H-1

HOW TO MAKE EXPEDIENT DEVICES H-1

WEAK POINTS OF ARMORED VEHICLES H-7

HOW TO MAKE EXPEDIENT DEVICES

In order to construct some of these devices, you must know how to prime charges electrically and nonelectrically. (app B).

FLAME DEVICES

These devices are used to obscure the vision of a vehicle's crew and to set the vehicle afire. The burning vehicle creates smoke and heat that will asphyxiate and burn the crew if they do not abandon the vehicle.

Molotov cocktail. This is made with a breakable container, a gas and oil mixture, and a cloth wick. To construct it, fill the container (usually a bottle) with the gas and oil mixture, and then insert the cloth wick into the container. The wick must extend both into the mixture and out of the container. Light the wick before throwing the Molotov cocktail. When the container hits a vehicle and breaks, the mixture will ignite, burning both the vehicle and the personnel around it.

Eagle fireball. This is made with an ammunition can, a gas and oil mixture, a white phosphorous grenade wrapped with detonating cord, tape, a nonelectric blasting cap, a fuse igniter, and a grapnel (or rope with bent nails). To construct an eagle fireball, fill the ammunition can with the gas and oil mixture. Wrap the grenade with detonating cord and attach a nonelectric firing system (app B) to the end of the detonating cord. Place the grenade inside the can with the time fuse extending out of it. Make a slot in the can's lid for the time fuse to pass through when the lid is closed. If available, attach a rope with bent nails or a grapnel to the can. When you throw the can onto a vehicle, the bent nails or the grapnel will help hold the can on the vehicle. Before throwing the can, fire the fuse igniter.

EAGLE FIREBALL

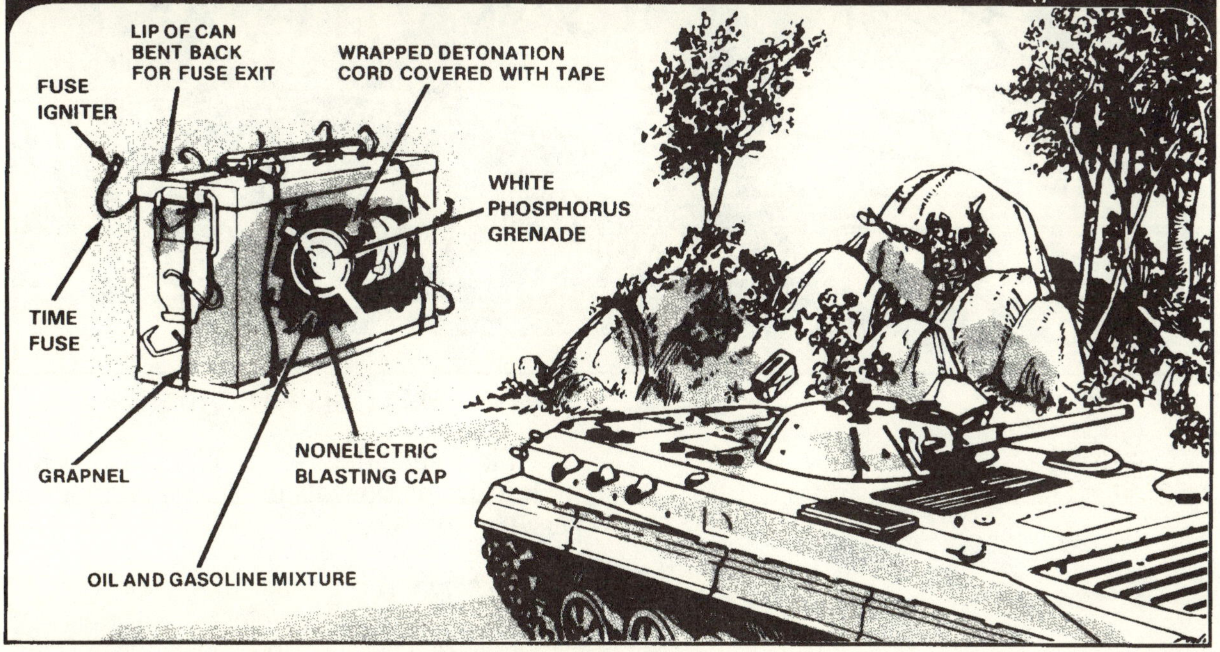

Eagle cocktail. This is made of a plastic or rubberized bag (a waterproof bag, a sandbag lined with a poncho, or a battery case placed inside a sandbag), a gas and oil mixture, a smoke grenade, a thermite grenade, tape, string, and communications wire or cord. To construct an eagle cocktail, fill the bag with the gas and oil mixture. Seal the bag by twisting its end and then taping or tying it. Attach the thermite and smoke grenades to the bag using tape, string, or communications wire. When attaching the grenades, do not bind the safety levers on the grenades. Tie a piece of string or cord to the safety pins of the grenades. Before throwing the eagle cocktail, pull the safety pins in both grenades.

EAGLE COCKTAIL

EXPLOSIVE DEVICES

Place such devices at vulnerable points to destroy components of tanks and armored personnel carriers.

Towed charge. This is made of rope or communications wire, mines or blocks of explosives, electrical blasting caps, tape, and electrical firing wire. To construct a towed charge, link a series of armed antitank mines together with rope or communications wire. If mines are not available, use about 25 to 50 pounds of explosives attached on a board (sled charge). Anchor one end of the rope on one side of a road and run its other end across the road to a safe position from which the charge may be pulled onto the road. Attach an electric firing system (app B) to each mine (or to the explosive on the sled charge) and connect those systems to the firing wire. Tape the firing wire to the rope running to the position from which the charge is pulled onto the road. At that position, conduct a circuit check (app B) and then connect the firing wire to a blasting machine. Just before a vehicle reaches the site of the towed charge, pull the charge onto the road so that it will be run over by the vehicle. When the vehicle is over it, fire the charge.

TOWED CHARGE

Pole charge. This is made of explosives (TNT or C4), nonelectric blasting caps, time fuse, detonating cord, tape, string or wire, fuse igniters, and a pole that is long enough for the mission. Prime the desired amount of explosives with two nonelectric firing systems, and attach the explosives to a board or some other flat material. The amount of explosives you use depends on the target to be destroyed. Tie or tape the board with the explosives to the pole. The time fuse should only be about 6 inches long. Before putting a pole charge on a target, fire the fuse igniters. Some good places to put a pole charge on a vehicle are under the turret, over the engine compartment, in the suspension system, and in the main gun tube (if the charge is made small enough to fit in the tube).

POLE CHARGE

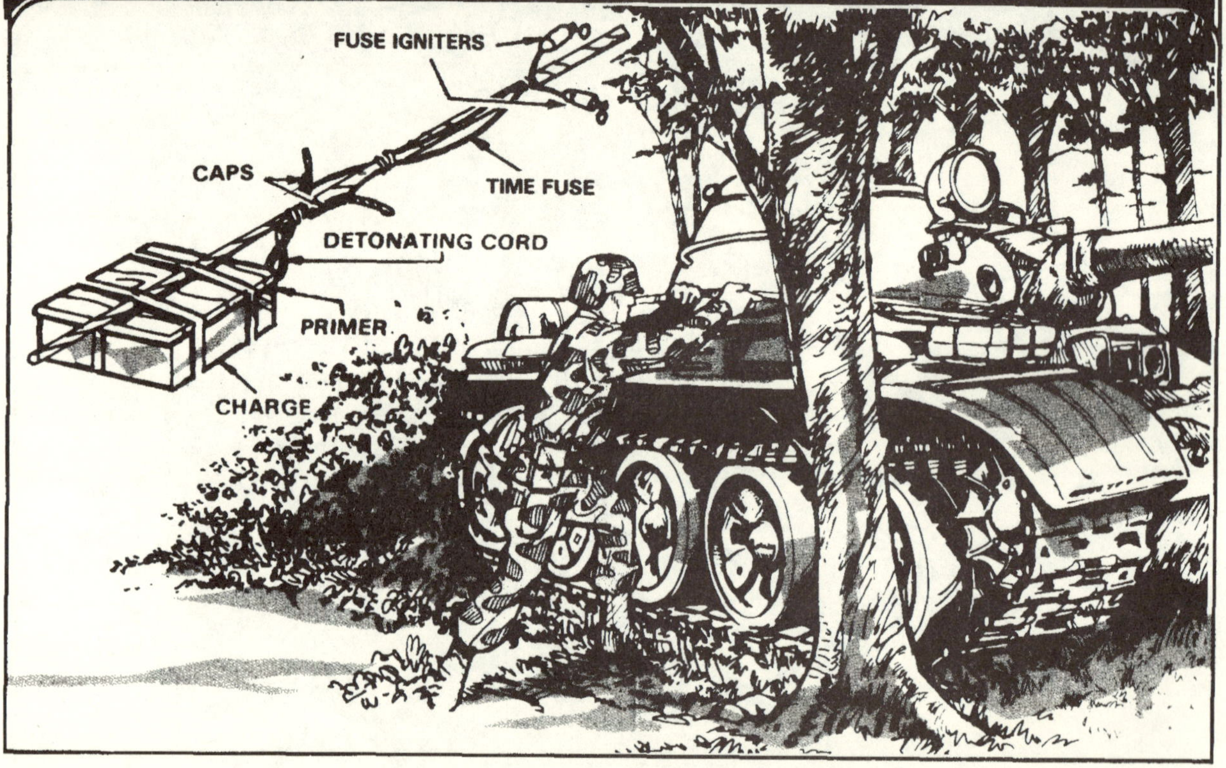

Satchel charge. This is made of explosives (TNT or C4), nonelectric blasting caps, time fuse, detonating cord, tape, fuse igniters, and some type of satchel. The satchel can be an empty sandbag, or demolitions bag, or some other material. To construct a satchel charge, fill the satchel with the amount of explosives needed for the mission. Prime the explosives with two nonelectric firing systems. Use only about 6 inches of time fuse. Seal the satchel with string, rope, or tape, and leave the time fuse and fuse igniter hanging out of the satchel. Before throwing a satchel charge onto a target, fire the fuse igniters.

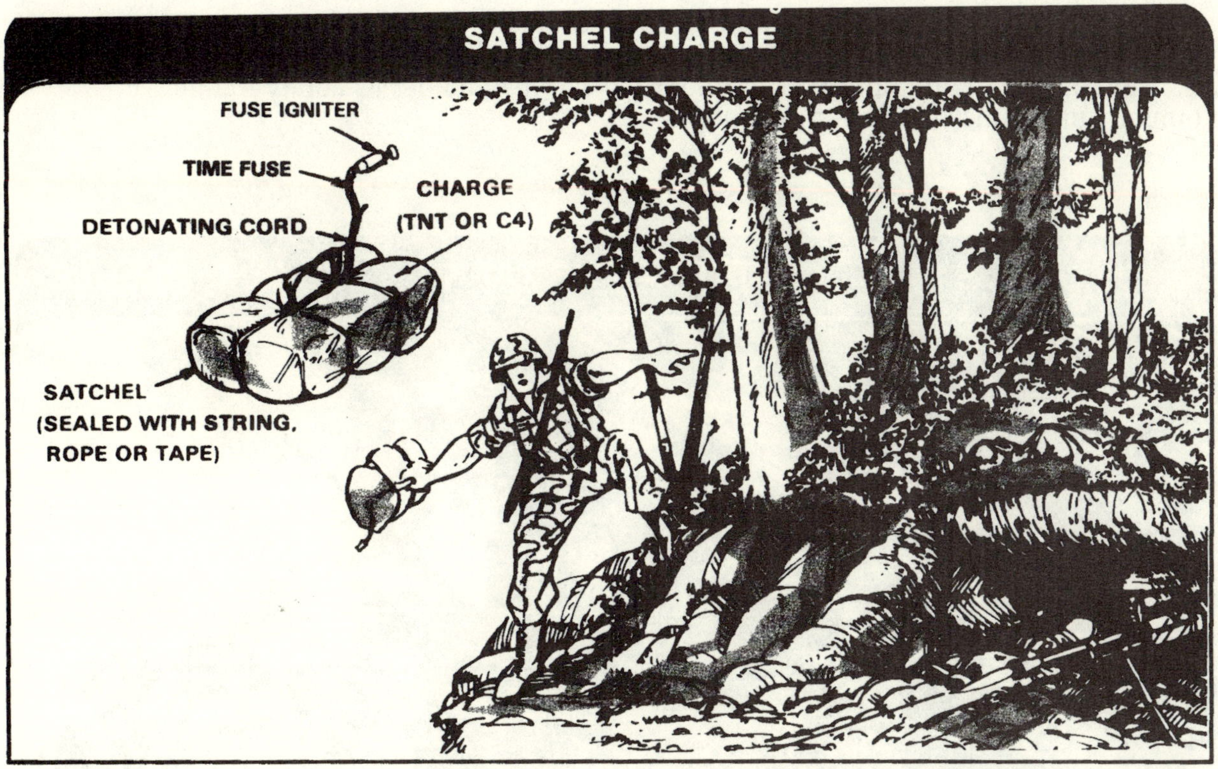

WEAK POINTS OF ARMORED VEHICLES

To use expedient devices successfully, you must know where the weak points of armored vehicles are. Some of the common weak points of armored vehicles are as follows:

- The suspension system.
- The fuel tanks (especially the external tanks).
- The ammunition storage compartments.
- The engine compartment.
- The turret ring.
- The armor on the sides, top, and rear (normally not as thick as that on the front).

NOTE: THE LOCATION OF SOME POINTS MAY VARY FROM VEHICLE TO VEHICLE.

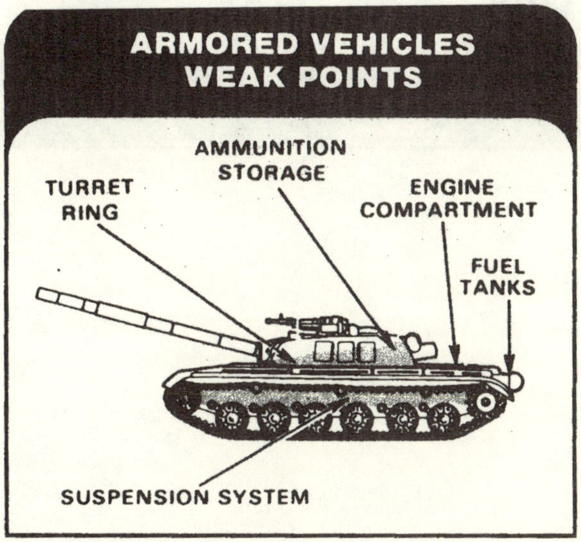

If a vehicle is "buttoned up" and you have no antiarmor weapons, fire your rifle at the vision blocks, at any optical equipment mounted outside the vehicle, into the engine compartment, at any external fuel tanks, or at the hatches. That will not destroy the vehicle, but may hinder its ability to fight.

BUTTONED-UP TANK

APPENDIX I

Range Cards

GENERAL

A range card is a rough sketch of the terrain around a weapon position. In the defense, you prepare a range card for the squad automatic weapon, the M60 and caliber .50 machine guns, and the Dragon, TOW, 106-mm RCLR, 90-mm RCLR, and LAW antiarmor weapons systems.

RANGE CARD DATA

A range card depicts **the following:**

- **Sectors of fire.**
- **A final protective line (FPL) or principal direction of fire (PDF).**
- **Targets and ranges to them.**
- **Prominent terrain features.**
- **Weapons symbols.**
- **Marginal data.**

Range cards for antiarmor weapons use target reference points (TRP) instead of FPL and PDF.

CONTENTS

GENERAL I-1
RANGE CARD DATA I-1
PREPARATION OF AN M60 MACHINE GUN RANGE CARD I-3
RANGE CARD FOR THE CALIBER .50 ....I-10
ANTIARMOR RANGE CARD I-10

SECTORS OF FIRE

Each gun is given a primary and a secondary sector of fire. Fire into your secondary sector of fire only if there are no targets in your primary sector, or if ordered to fire there. Your gun's primary sector includes an FPL, a PDF, or a TRP.

FINAL PROTECTIVE LINE

Where terrain allows, your leader assigns an FPL to your weapon. The FPL is a line on which you shoot grazing fire across your unit's front.

The FPL will be assigned to you only if your leader determines there is a good distance of grazing fire. If there is, the FPL will then dictate the location of the primary sector. The FPL will become the primary sector limit (right or left) closest to friendly troops. When not firing at other targets, you will lay your gun on the FPL or PDF.

DEAD SPACE

Dead space is an area that direct fire weapons cannot hit. The area behind houses and hills or within orchards, for example, is dead space.

PRINCIPAL DIRECTION OF FIRE

When the terrain does not lend itself to an FPL, your leader will assign a PDF instead. The direction should be toward a gully or down a ditch that leads into your position. The gun is positioned to fire directly down this approach rather than across the platoon's front.

TARGETS

Your leader may also designate locations within your sector of fire where targets are most likely to appear. These locations should be noted on your range card.

TARGET REFERENCE POINTS

Target reference points are natural or manmade features within your sector that can be used for quick location of targets. Target reference points are used primarily for controlling DIRECT FIRE weapons only; however, TRP should appear on the company target list.

MAXIMUM ENGAGEMENT LINE

The maximum engagement line is a line beyond which you cannot engage a target. This line may be closer than the maximum engagement range of your weapon. Both the terrain and the maximum engagement range of your weapon will determine the path of the maximum engagement line. *This line is used for antiarmor range cards.*

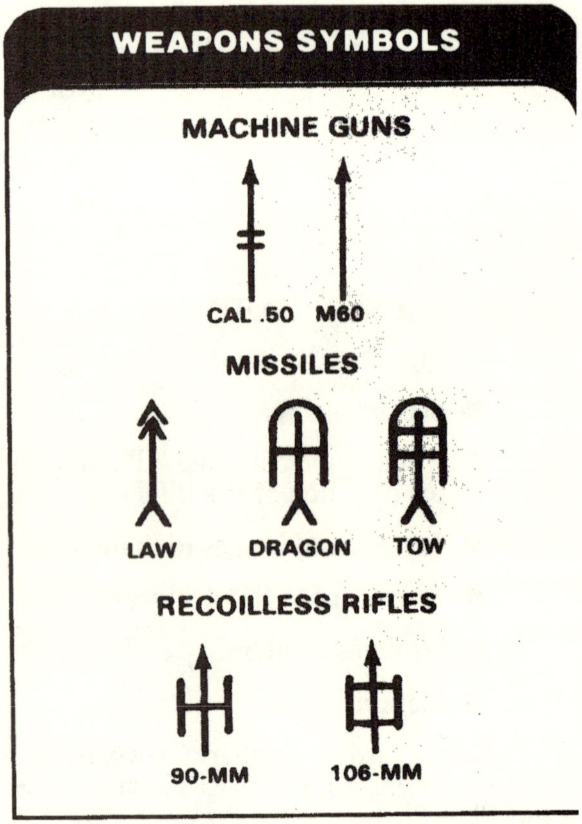

PREPARATION OF AN M60 MACHINE GUN RANGE CARD

Range cards are prepared immediately upon arrival in your position. **To prepare an M60 machine gun range card:**

- **Orient the card so that both the primary and secondary sectors of fire (if assigned) can fit on it.**
- **Draw a rough sketch of the terrain to the front of your position. Include any prominent natural and manmade features which could be likely targets.**
- **Draw your position at the bottom of the sketch. Do not put in the weapon symbol at this time.**
- **Fill in the marginal data to include:**
  - ☐ **Gun number (or squad).**
  - ☐ **Unit (only platoon and company). Date.**
  - ☐ **Magnetic north arrow.**

EXAMPLE OF THE RANGE CARD

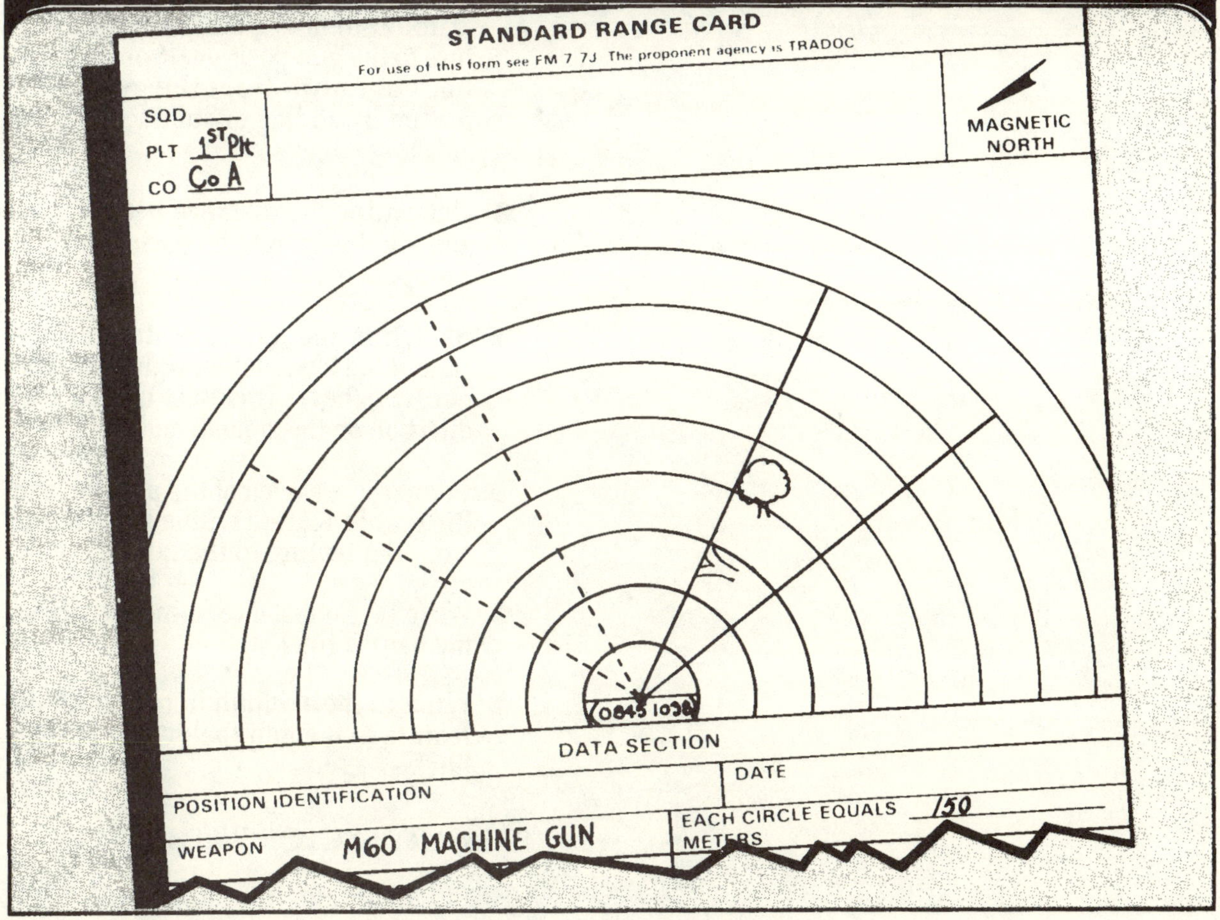

- Use the lensatic compass to determine magnetic north and sketch in the magnetic north arrow on the card with its base starting at the top of the marginal data section.

- Determine the location of your gun position in relation to a prominent terrain feature, such as a hilltop, road junction or building. If no feature exists, place the eight-digit map coordinates of your position near the point where you determined your gun position to be. If there is a prominent terrain feature within 1,000 meters of the gun, use that feature. Do not sketch in the gun symbol at this time.

- Using your compass, determine the azimuth in mils from the terrain feature to the gun position. (Compute the back azimuth from the gun to the feature by adding or subtracting 3,200 mils.)

- Determine the distance between the gun and the feature by pacing or from a map.

- Sketch in the terrain feature on the card in the lower left or right hand corner (whichever is closest to its actual direction on the ground) and identify it.

- Connect the sketch of the position and the terrain feature with a barbed line from the feature to the gun.

- Write in the distance in meters (above the barbed line).

- Write in the azimuth in mils from the feature to the gun (below the barbed line).

PRIMARY SECTOR WITH AN FPL

To add an FPL to **your range card:**

- Sketch in the limits of the primary sector of fire as assigned by your leader.
- Sketch in the FPL on your sector limit as assigned.
- Determine dead space on the FPL by having your buddy walk the FPL. Watch him walk down the line and mark spaces which cannot be grazed.
- Sketch dead space by showing a break in the symbol for an FPL, and write in the range to the beginning and end of the dead space.
- Label all targets in your primary sector in order of priority. The FPL is number one.

PRIMARY SECTOR WITH AN FPL

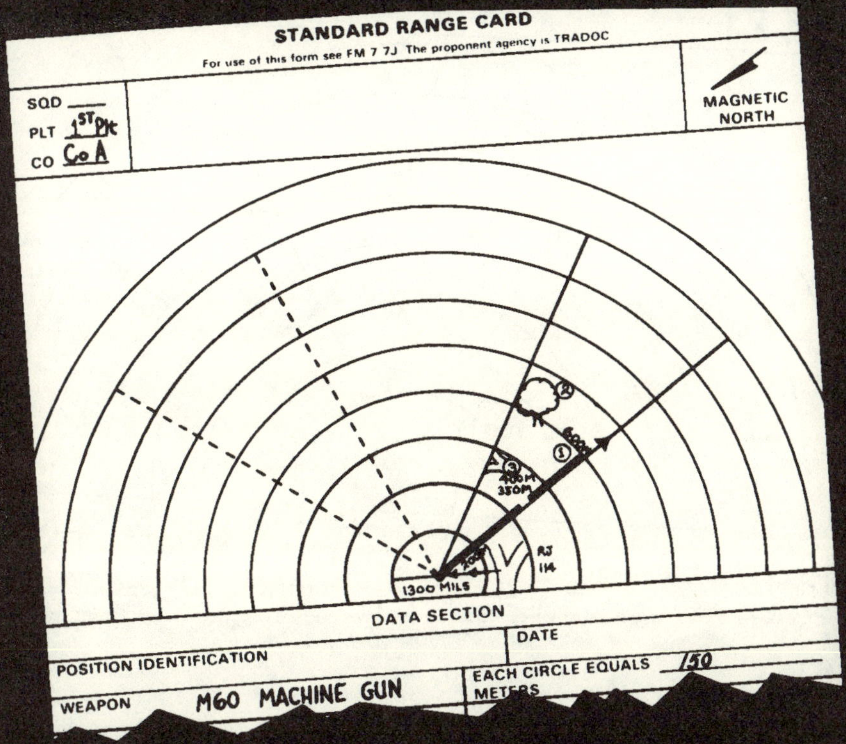

PRIMARY SECTOR WITH A PDF

To prepare your range card when assigned a **PDF instead of an FPL:**

- **Sketch in the limits of the primary sector of fire as assigned by your leader (sector should not exceed 875 mils, the maximum traverse of the tripod-mounted M60).**
- **Sketch in the symbol for an automatic weapon oriented on the most dangerous target within your sector (as designated by your leader). The PDF will be target number one in your sector. All other targets will be numbered in priority.**
- **Sketch in your secondary sector of fire (as assigned) and label targets within the secondary sector with the range in meters from your gun to each target. Use the bipod when it is necessary to fire into your secondary sector. The secondary sector is drawn using a broken line. Sketch in aiming stakes, if used.**

COMPLETE DIAGRAM WITH PDF

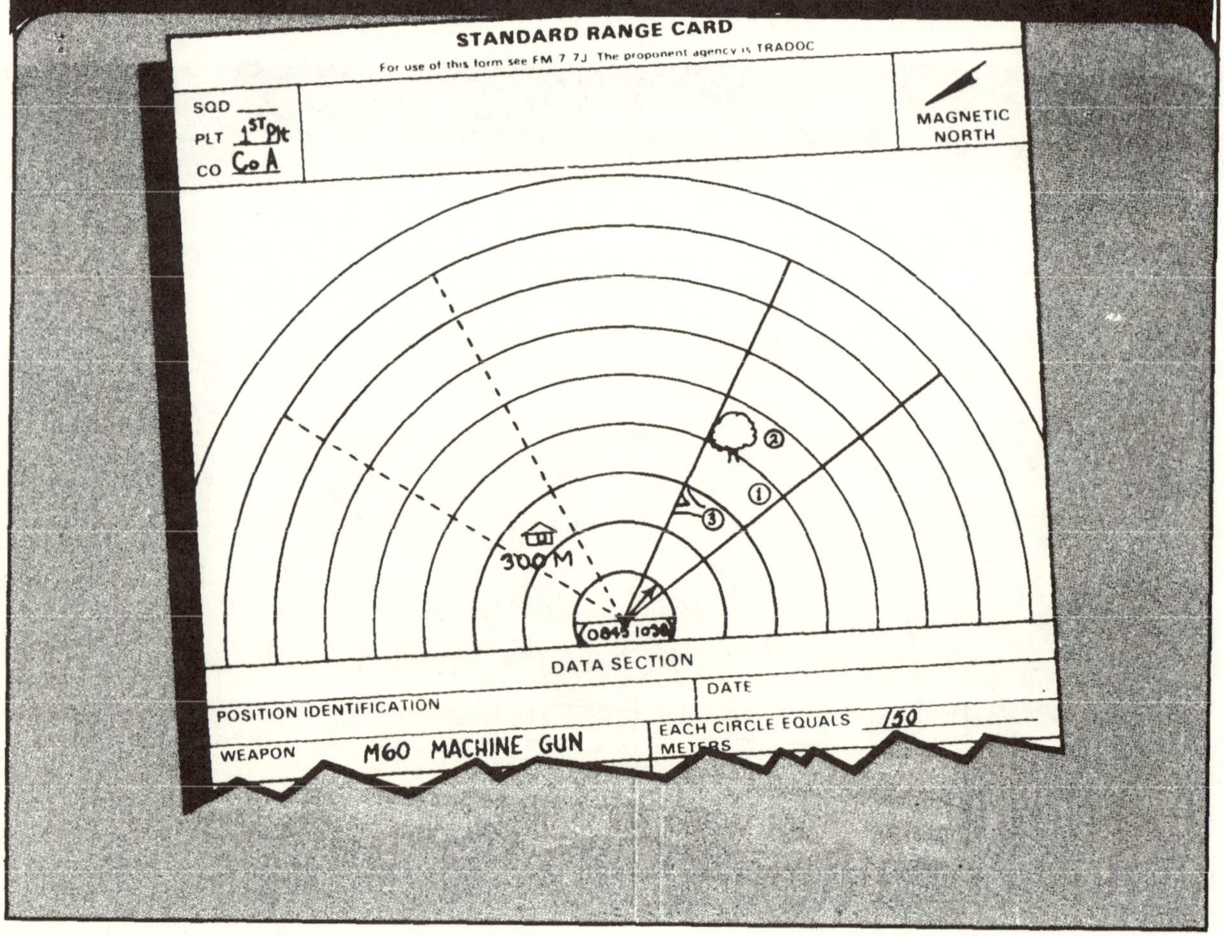

COMPLETE DIAGRAM WITH FPL

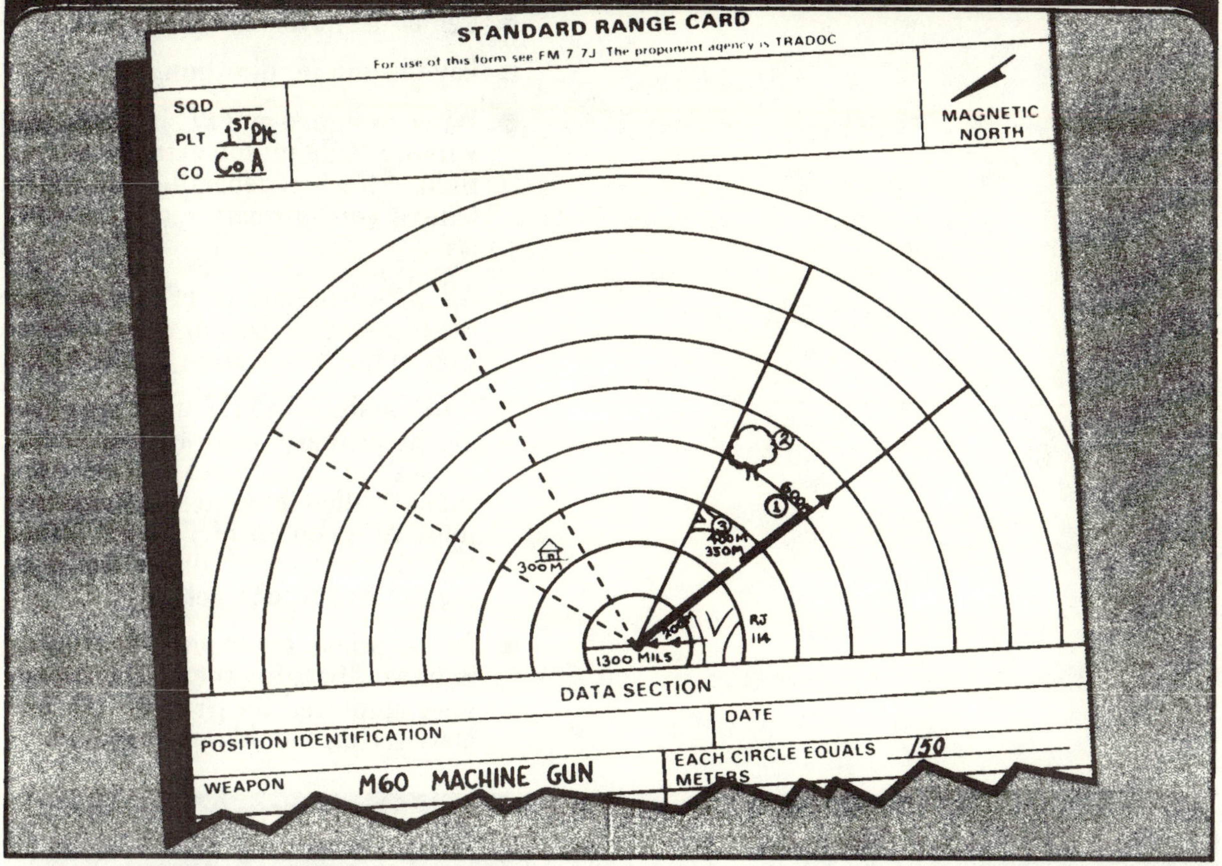

DATA SECTION OF M60 RANGE CARD

The data section of the range card lists the data necessary to engage targets identified in the sketch. The sketch does not have to be to scale, but the data must be accurate. The data section of the card can be placed on the reverse side or below the sketch if there is room. Draw a data section block (if you do not have a printed card), with the **following items:**

DATA SECTION

NO	DIRECTION	ELEVATION	RANGE	DESCRIPTION	REMARKS
1		+50/3	600	FPL	-4
2	R 105	+50/40	500	LONE PINE	
3	L 235	0/28	350	TRAIL JUNCTION	W15/L7

To prepare the data section of **the M60** range card:

- Center the traversing handwheel.
- Lay the gun for direction.
- When assigned an FPL, lock the traversing slide on the extreme left or right of the bar, depending on which side of your primary sector the FPL is on.
- Align the barrel on the FPL by moving the tripod legs. (Do not enter a direction in the data section for the FPL.)
- When assigned a PDF, align your gun on the primary sector by traversing the slide to one side and then move the tripod to align the barrel on your sector limit. Align on the PDF by traversing the slide until your gun is aimed at the center of the target.
- Fix the tripod legs in place by digging in or sandbagging them. The tripod, once emplaced for fire into the primary sector, should not be moved.

TRAVERSING AND ELEVATING MECHANISM

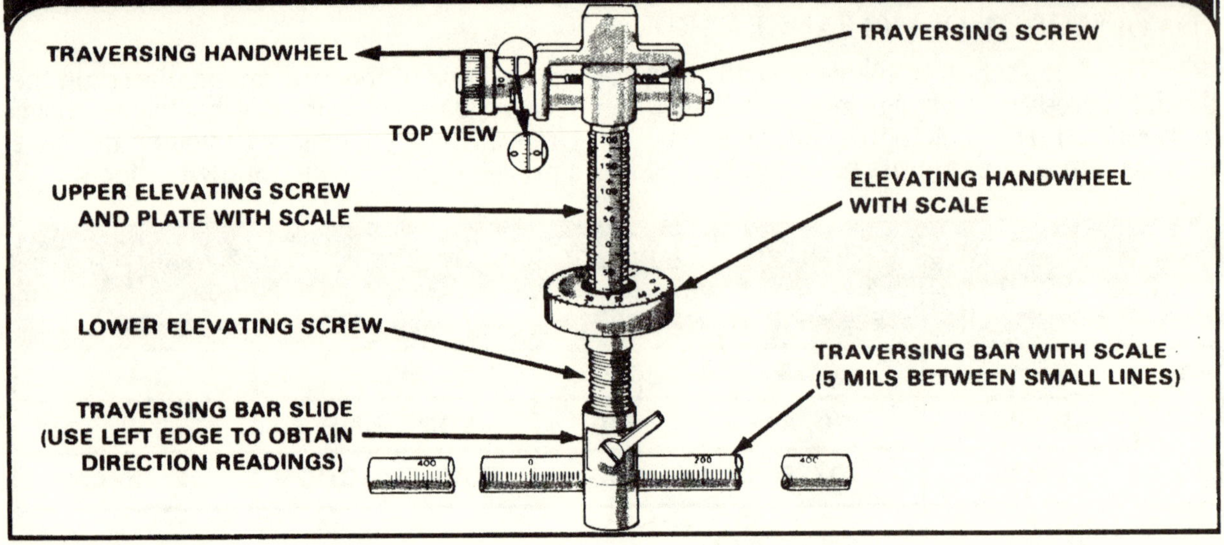

To read the direction to **each target:**

- **Lay your gun on the center of the target.**
- **Read the direction directly off the traversing bar at the left edge of the traversing bar slide.**
- **Enter the reading under the direction column of your range card data section.**

A reading of left or right is determined by the direction of your barrel (just the opposite of the slide).

To read elevation for **your targets:**

- **Lay your gun on the base of the target by rotating the elevating handwheel.**
- **Read the number (to include a plus or minus sign, except for "0") *above* the first visible line on the elevating scale.**
- **The sketch reads –50.**
- **Read the number on the elevating handwheel that is in line with the indicator.**
- **The sketch reads 3.**
- **Enter this reading under the ELEVATION column of your range card data section, separating the two numbers with a slash (/). Always enter the reading from the upper elevating bar first. (The sketch reads –50/3.)**

Enter range to each target under the appropriate column in the data section.

Enter the description of each target under the appropriate column in the data section.

REMARKS COLUMN

Fill in the remarks column for each target **as needed:**

- **Enter the width and depth (in mils) of linear targets. The —4 in the illustration indicates that by depressing your barrel 4 mils the strike of your rounds will go down to ground level along the FPL.**
- **When entering the width of the target, be sure to give the width in mils and express it as two values. For instance, the illustration shows that target number 3 has a width of 15 mils. The second value, L7, means that once the gun is laid on your target, traversing 7 mile to the LEFT will lay the gun on the left edge of the target.**
- **Enter *aiming stake* if one is used for the target.**
- **No data for the secondary sector will be determined since your gun will be fired in the bipod role.**

COMPLETED DATA SECTION

DATA SECTION

WEAPON: 1
UNIT: 1ST PLT/CoA
DATE:

MAGNETIC NORTH

EACH CIRCLE: 150 METERS

NO.	DIRECTION	ELEVATION	RANGE	DESCRIPTION	REMARKS
1		–50/3	600	FPL	–4
2	R105	+50/40	500	LONE PINE	
3	L235	0/28	350	TRAIL JUNCTION	W15/L7

RANGE CARD FOR THE CALIBER .50

The only differences between an M60 and a caliber .50 range card are **as follows:**

- **The machine gun symbol is different.**
- **There are 800 mils of traverse with the caliber .50 compared to 875 mils with the M60.**
- **Maximum grazing fire with the caliber .50 is 1,000 meters compared to 600 meters with the M60.**
- **The caliber .50 machine gun has a secondary sector of fire; but it must be marked by aiming stakes since the caliber .50 machine gun has no bipod.**

ANTIARMOR RANGE CARD

The purpose of an antiarmor range card is to show a sketch of the terrain that a weapon has been assigned to cover by fire. Range cards for 90-mm RCLR, 106-mm RCLR, Dragon, and TOW are all prepared the same. By using a range card, you can quickly and accurately determine the information needed to engage targets in your assigned sector. Before you prepare a range card, your leader will show you where to position your weapon so you can best cover your assigned sector of fire. He will then, **again,** point out the terrain you are to cover. He will do this by assigning you a sector of fire or by assigning left or right limits indicated by either terrain features or azimuths. If necessary, he may also assign you more than one sector of fire and will designate the sectors as primary and secondary.

PREPARATION OF THE RANGE CARD

Once you have all the necessary information, you can begin preparing your range card, depending upon the priority of other jobs you must perform (such as preparing and camouflaging your firing position). If you are assigned alternate or supplementary firing positions, a range card is required for them also.

PROCEDURES

In the lower center of your range card, indicate your firing position by drawing the symbol for your assigned weapon. Also indicate the direction of magnetic north (not a requirement for a LAW).

Draw and label your sector sketch. Draw roads, bridges, buildings, streams, hills, and woods. Be as accurate as you can.

ANTIARMOR RANGE CARD SHOWING POSITION

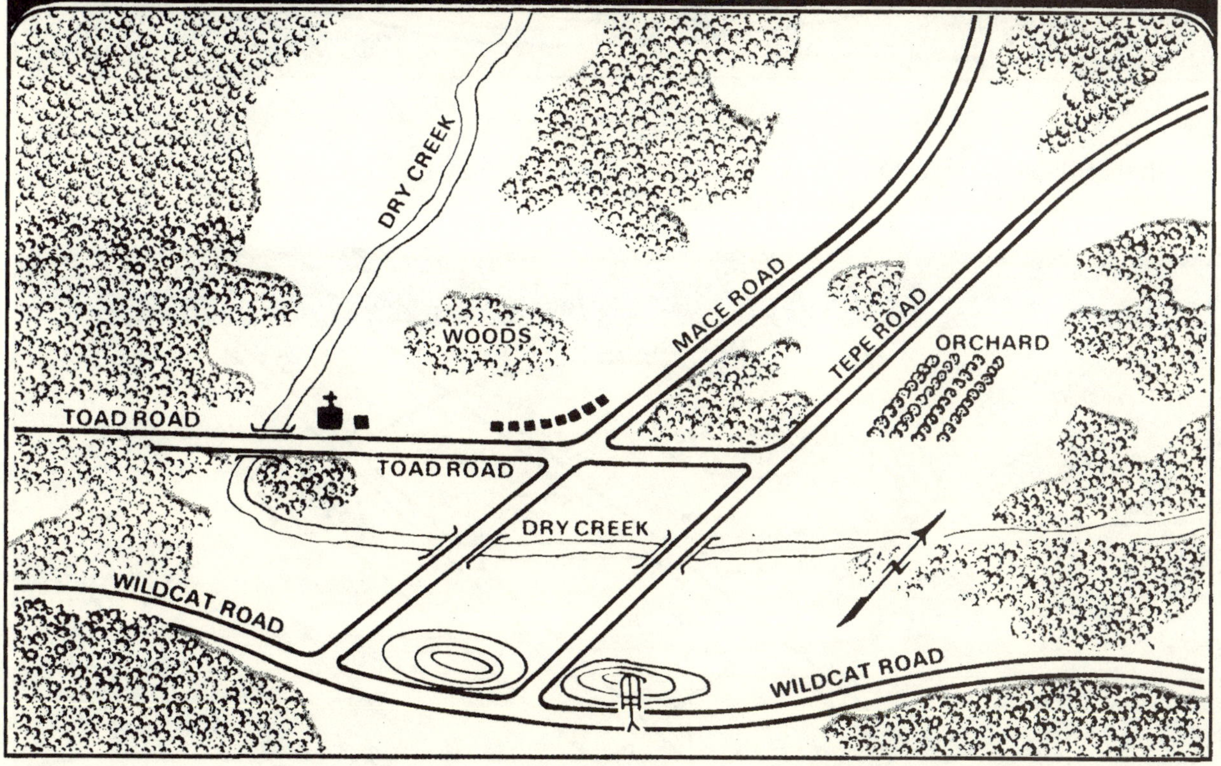

Show the location to your firing positions by drawing an arrow from a nearby recognizable terrain feature and assign it number one. Add the azimuth and distance from the terrain feature to your firing position. (This is not a requirement for a LAW.)

INDICATED AZIMUTH FROM TERRAIN FEATURE

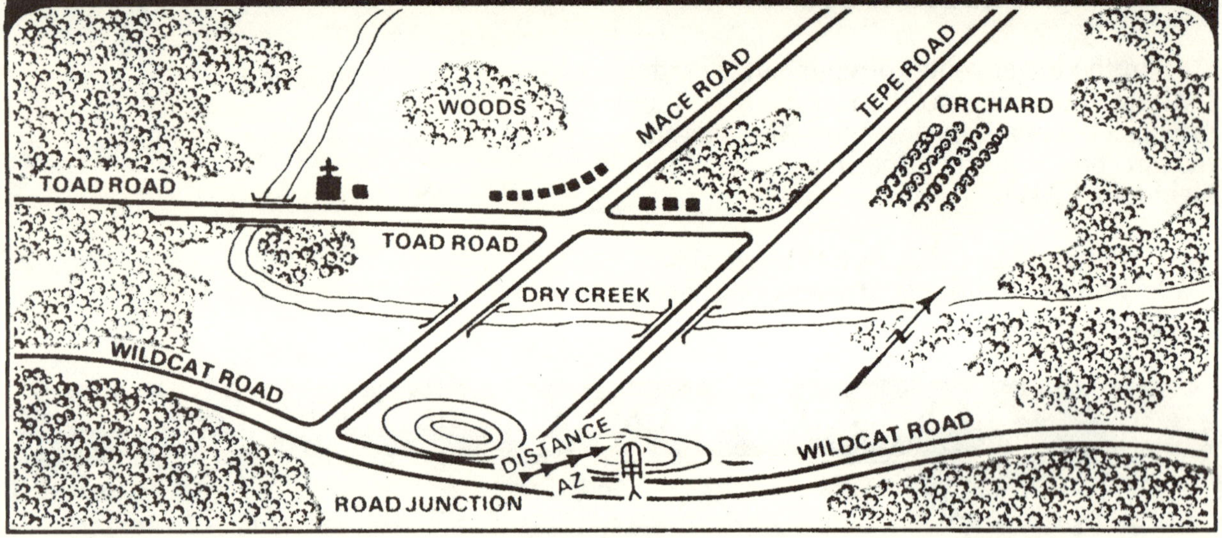

Now draw your sector. This is an enclosed line that outlines your sector of fire. The *maximum engagement line* is a segment of the sector line and indicates the maximum range that targets may be engaged.

SECTOR OF FIRE

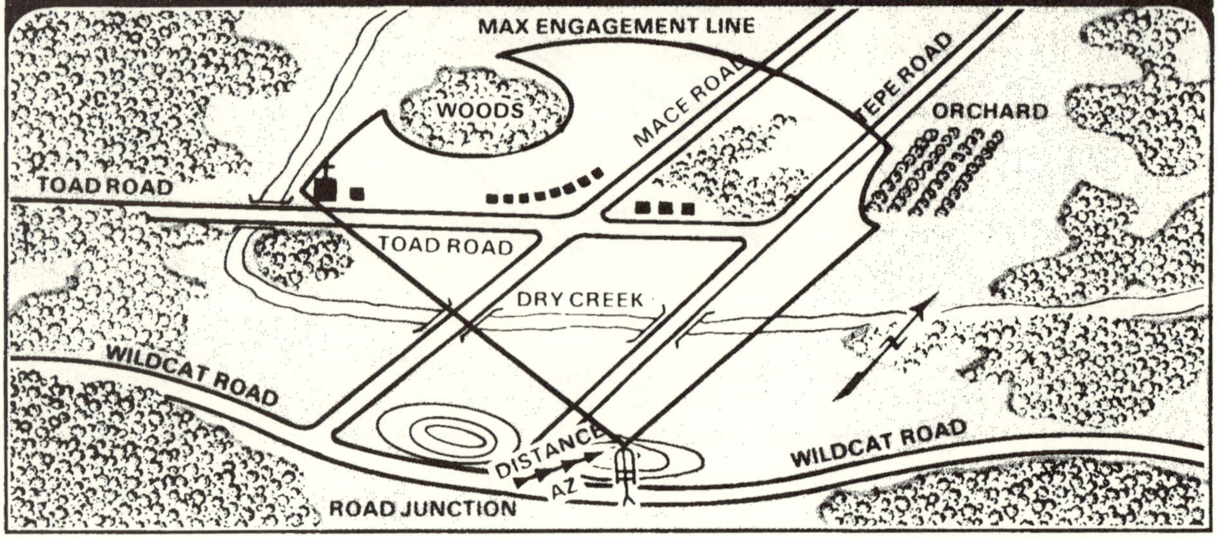

Draw in the dead space in your sector. Be sure to indicate by an enclosed line those areas you cannot hit. Remember, your sector of fire can be any shape and size.

DEAD SPACE

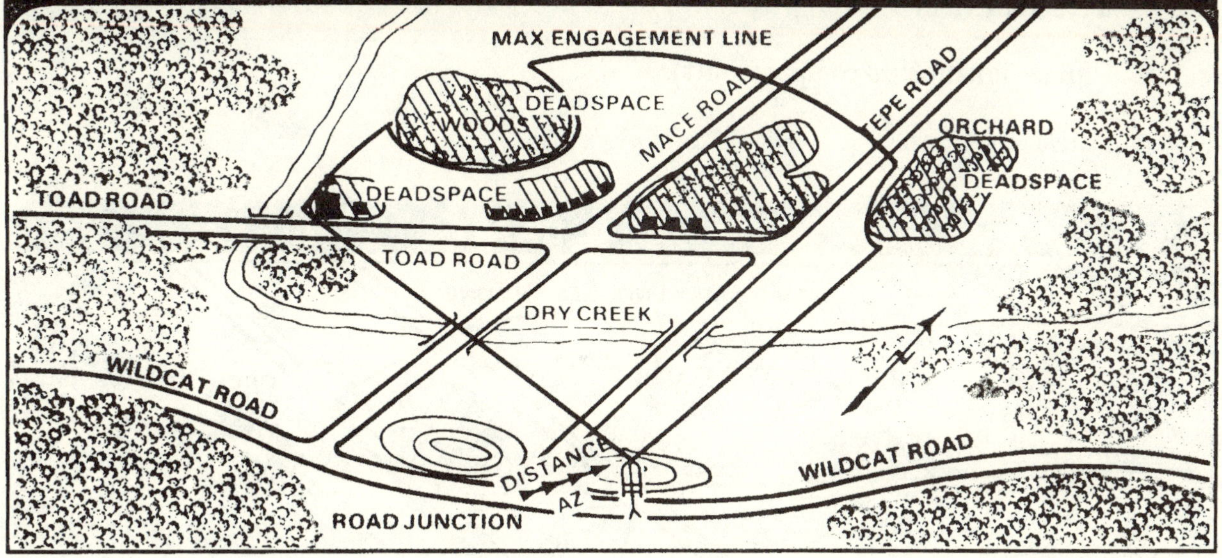

Next, draw in the range and azimuths to expected target engagement locations and TRPs in your sector. (Azimuth is not a requirement for a LAW.)

INDICATE RANGE AND AZIMUTH

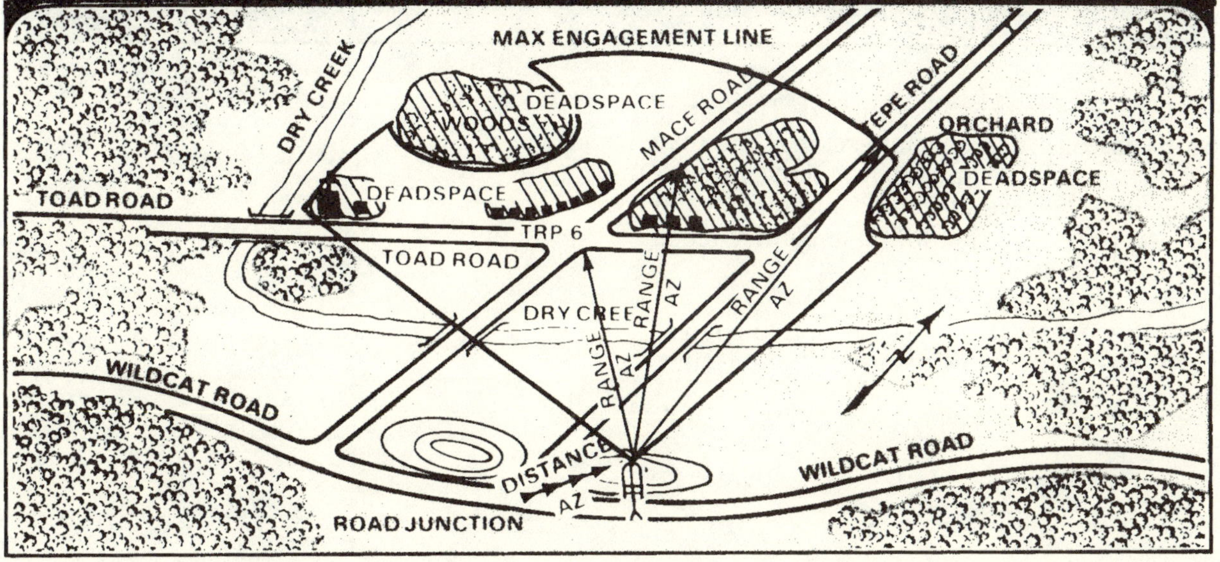

Write in marginal data. Marginal data **must include the following:**

- **Type position (primary, supplementary, or alternate).**
- **Unit designation (to company only).**
- **Date/Time group.**

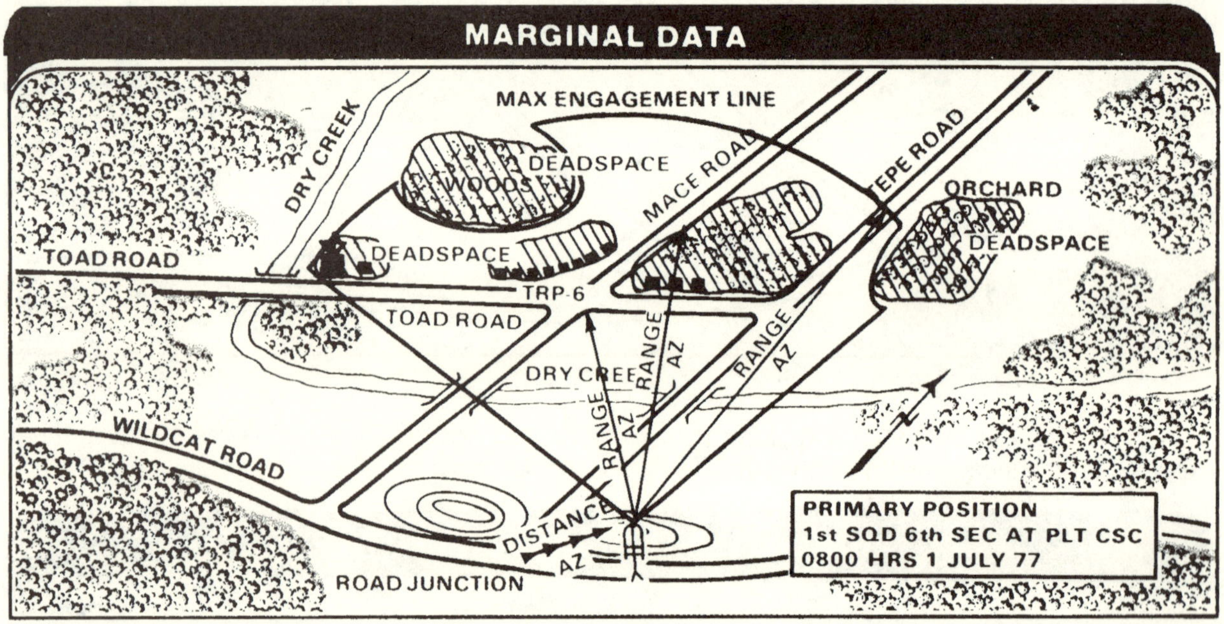

Your range card is finished. The range card you construct for your sector of fire may not look exactly like those shown in this manual. Remember, however, the basic information and method of construction for all antiarmor range cards are the same.

Prepare your range card in two copies, Keep one copy at the weapon and send the other to your leader.

GLOSSARY

Acronyms And Abbreviations

A

AM	amplitude modulation
ammo	ammunition
AR	automatic rifle
ATGM	antitank guided missile

c

CB	chemical-biological
CEOI	communications-electronics operation instructions
cm	centimeter

E

EMP	electromagnetic pulse

F

FM	frequency modulation
FPL	final protective line
ft	foot/feet

G

GL	grenade launcher

H

HEDP	high explosive dual purpose
ht	height

I

in	inch

K

kg	kilogram
km	kilometer

L

LAW	light antitank weapon
LCE	load-carrying equipment

M

m	meter
mg	machine gun
min	minute
MOPP	mission-oriented protective posture
MOS	military occupational specialty

N

NBC	nuclear, biological, and chemical

O

OP	observation post
oz	ounce

P

PDF	principal direction of fire
PW	prisoner of war
pwr	power

R

RCLR	recoilless rifle
rd	round
RL	rocket launcher

s

SOP	standing operating procedure
SUT	small unit leader

T

TNT	chemical composition similar to dynamite
TRP	target reference point

U

US	United States

W

wt	weight

References

REQUIRED PUBLICATIONS

Required publications are sources which users must read in order to understand or to comply with FM 21-75.

NONE

RELATED PUBLICATIONS

Related publications are sources of additional information. Users do not have to read them in order to understand FM 21-75.

FIELD MANUAL (FM)

FM 3-12	Operational Aspects of Radiological Defense
FM 5-25	Explosives and Demolitions
FM 21-11	First Aid for Soldiers
FM 21-40	NBC (Nuclear, Biological, and Chemical) Defense
FM 21-60	Visual Signals
FM 21-76	Survival, Evasion, and Escape
FM 23-9	M16A1 Rifle and Rifle Marksmanship
FM 23-11	90-mm Recoilless Rifle, M67
FM 23-31	40-mm Grenade Launchers M203 and M79
FM 23-35	Pistols and Revolvers
FM 23-65	Browning Machinegun, Caliber .50 HB, M2
FM 23-67	Machinegun, 7.62-mm, M60
FM 90-10-1	An Infantryman's Guide to Urban Combat

TECHNICAL MANUAL (TM)

TM 9-1345-203-128P	Operator's and Organizational Maintenance Manual Land Mines

GRAPHIC TRAINING AID (GTA)

GTA 5-10-27	Mine Card

Index

	Page
Aiming and Sector Stakes	2-13
Bangalore Torpedo	C-7
Biological Weapons	5-5
decontamination	5-12
Blast from Nuclear Explosions	5-1, 5-2
Blistering Agents	5-10
Bleeding, Stop the	8-9
Blood Agents	5-10
Caliber .50 Machine Gun	G-9
Camouflage	1-3
camouflage considerations	1-3
colors used in camouflage	1-10
fighting positions	1-8
how to camouflage	1-7
Check for and Restore Breathing	8-4
Check for and Restore Heartbeat	8-6
Chemical Agent Decontamination	5-11
Chemical Agent Detection Paper (ABC-M8)	5-7
Chemical and Biological Weapons	5-5
alarms	5-7
characteristics	5-5
chemical attacks	5-9
detection	5-5
effects	5-5
protection against	5-8
Choking Agents	5-10
Classes of Fire	G-14
Clear the Airways	8-3

	Page
Colors Used in Camouflage	1-10
Combat Intelligence, and Counterintelligence	6-1
counterintelligence measures	6-5
sources of information	6-2
what to report	6-2
Communications	7-1
communications security	7-10
means of communication	7-1
radio equipment	7-11
radiotelephone procedures	7-4
wire and telephone equipment	7-15
Concealment	1-3
for fighting positions	2-2
Counterintelligence	6-1
Cover	1-1
fighting positions	2-2
types	1-2
Covering the Target Area	G-18
Crawls	3-3
Curved Hole Fighting Positions	2-8
Demolitions, How to Prepare Firing Systems	B-1
electric	B-2
nonelectric	B-6
Dragon Fighting Positions	2-26
Dress and Bandage the Wound	8-15
Elbow Holes	2-12
Electric Demolition Firing System Preparation	B-6
EMP (Electro Magnetic Pulse)	5-2, 5-3
Entering a Building	D-7
high level entries	D-7
how to use hand grenades	D-10
low level entries	D-9
roof landing	D-8
Escape	F-2
Evasion	F-2

Page

Field Expedient Antiarmor Devices H-1
how to make expedient devices H-1
weak points of armored vehicles H-7

Fields of Fire 2-6

Fighting Positions 2-1
camouflaging 2-18
compartment 2-18
concealment of 2-5
cover 2-2
depth 2-11
hasty 2-8
how to build 2-8
how to camouflage 1-8
in urban areas D-10
loophole positions D-15
one-man 2-19
sectors and fields of fire 2-5
storage compartment 2-31
trenches 2-30
two-man 2-30
See Also: type of weapon

Fire, Characteristics of G-11

Fire, Classes of G-14

Fire Control G-19

Fire Distribution G-17

Fire and Movement 3-9

Fire Team Wedge 3-8

Firing Ports 2-11

Firing Systems, Demolitions B-1

First Aid and Personal Hygiene 8-1
do's and don'ts of first aid 8-16
lifesaving measures 8-2
personal hygiene 8-16

Flame Assault Shoulder Weapon Fighting Position (FLASH) 2-29

Glossary Glossary-1

Grenade Sumps 2-14, 2-22

	Page
Hasty Fighting Positions	2-8
Immediate Actions While Moving	3-5
Light Antitank Weapons (M72A2) (LAW)	G-6
fighting positions	2-29
Lifesaving Measures	8-2
Loophole Fighting Positions	D-15
Luminous Night Identification Device	3-9
M8 Automatic Chemical-Agent Alarm	5-6
M16A1 Rifle Characteristics	G-2
M1911A1 Pistol Characteristics	G-2
M47 Dragon Medium Antitank Weapon	G-8
M60 Machine Gun Characteristics	G-2
M60 Weatherproof Fuse Igniter	B-4
M67 90-mm Recoilless Rifle (RCLR)	G-11
fighting position	2-28
M72A2 Light Antitank Weapon (LAW)	G-6
fighting position	2-29
M202A1, Multishot Rocket Launcher 66-mm (FLASH)	G-7
M203 40-mm Grenade Launcher	G-4
M256 Chemical Agent Detector Kit	5-7
M258A1 (Chemical Agent Decontamination Kit)	5-11
Machine Gun Fighting Position	2-20
Messenger Communications	7-4
Minefields, How to Breach and Cross	C-2
Mines	A-1
M14, blast, antipersonnel	A-2
M15, metallic heavy antitank	A-22
M16A1, series bounding, antipersonnel	A-6
M18A1, fragmentation, antipersonnel	A-11
M21, antitank	A-27
M24, off-route antitank	A-32
M26, antipersonnel	A-17

MOPP (Mission-Oriented Protective Posture) Levels 5-9

Mouth-to-Mouth Resuscitation 8-5

Movement 3-1
fire and movement 3-9
immediate actions while moving 3-5
movement techniques 3-2
moving with stealth 3-4
moving with tanks 3-10
moving within a team 3-8

Movement in Urban Areas D-1
how to cross a wall D-2
how to cross open areas D-4
how to move around a corner D-2
how to move in a building D-6
how to move parallel to a building D-4
how to move past a window D-3
route selection D-5

Moving with Tanks 3-10

Moving within a Team 3-8

Nerve Agents 5-9

Night Observation 4-3

Nonelectric Demolition Firing System Preparation B-2

Nuclear, Biological, and Chemical Warfare (NBC) 5-1
nuclear weapons 5-2
chemical and biological weapons 5-5

Nuclear Radiation 5-2, 5-3

Nuclear Weapons 5-2
characteristics 5-2
effects 5-3
protection against nuclear attacks 2-4, 5-3

Observation 4-1
how to observe 4-2
range estimation 4-4
things to look and listen for 4-4

Obstacles C-1
how to breach and cross a minefield C-2
how to breach and cross wire obstacles C-5

Overhead Cover 2-14
camouflaging with sod 2-18

One-Man Fighting Position 2-19

Personal Hygiene 8-16

Phonetic Alphabet and Numbers 7-5

Pressure Dressing 8-11

Prevent Shock 8-14

Prisoners of War (PWs) 6-3

Protective Clothing and Equipment Against CB Attack 5-8

Prowords 7-7

Radio Communications 7-2

Radio Equipment 7-11
AN/PRC-9 (receiver) 7-13
AN/PRC-68 small unit transceiver (SUT) 7-12
AN/PRC-77 radio 7-11
AN/PRT-4 (transmitter) 7-11

Radiotelephone Procedure 7-4
numbers 7-6
phonetic alphabet 7-5
prowords 7-7

Range Cards I-1
antiarmor range card I-10
preparation of M60 machine gun range card I-3
range card for the caliber .50 I-10
sectors of fire I-2

Range Estimation 4-4

Recoilless Rifle, 90-mm, (RCLR) G-11
fighting position 2-28

Reel Equipment CE-11 7-17

Resistance F-2

Revetment 2-19

Roof Fighting Positions D-12 thru D-15

Rush 3-3

SALUTE Report 6-2

Sectors and Fields of Fire 2-5

Sector Stakes 2-13

Sound Communications 7-3

Stop the Bleeding 8-9

Storage Compartments 2-31

Sumps 2-14, 2-22

Survival, Evasion, Resistance, and Escape F-1

TA-1 Telephone 7-15

TA-1 Telephone 7-16

Thermal Radiation 5-2

Tourniquets 8-13

Tracking E-1
- camouflage E-8
- countertracking E-10
- displacement E-2
- fundamentals of tracking E-2
- interpretation/immediate use of intelligence E-8
- littering E-8
- side as unit of measure E-4
- staining E-6
- tracker dog E-10
- tracker qualities E-2
- tracking teams E-9
- weathering E-7

Trenches 2-30

Two-Man Fighting Position 2-8
- modified 2-25

Urban Areas D-1
- how to enter a building D-7
- how to move D-1
- how to use fighting positions D-10

Visual Communications 7-3

Wire Communications 7-4

Window Fighting Positions D-11

Wire Obstacles, How to Breach and Cross C-5

Wire and Telephone Equipment 7-15
reel equipment CE-11 7-17
wire-laying techniques 7-15

Weapons and Fire Control G-1
characteristics of fire G-11
covering the target area G-18
fire control G-19
fire distribution G-17
weapons G-2